68 Structure and Bonding

AF739290

Theoretical Approaches

With Contributions by
R. L. Johnston O. Kahn D. M. P. Mingos
O. S. Mortensen

With 80 Figures and 18 Tables

Springer-Verlag Berlin
Heidelberg GmbH

ISBN 978-3-662-15156-3 ISBN 978-3-540-47730-3 (eBook)
DOI 10.1007/978-3-540-47730-3

Library of Congress Cataloging-in-Publication Data. Theoretical approaches. (Structure and bonding; 68) Bibliography: p. Includes index. 1. Chemical bonds. 2. Chemical structure. I. Johnston, R. L. (Roy Luigi) II. Series. QD461.S92 vol. 68 541.2 87-26438

Originally published by Springer-Verlag Berlin Heidelberg New York in 1987.
Softcover reprint of the hardcover 1st edition 1987

Typesetting: Mitterweger Werksatz GmbH, 6831 Plankstadt, Germany

2152/3140-543210

Table of Contents

A Noncommuting-Generator Approach to Molecular Symmetry

O. Sonnich Mortensen

Fysisk Institut, Odense Universitet, DK-5230 Odense M, Denmark

A novel method of treating point group symmetry is presented. It is based on finding common eigenfunctions to sets of *non*-commuting operators, the operators being suitably chosen generators for the point groups. The character tables of the conventional method are replaced by eigenvalue tables, that are no more complex than character tables but allow straight forward calculation of matrixelements (Wigner-Eckart theorem). Half-integer spin is treated as simply as integer spin, without introducing the socalled double groups.

Structure and Bonding 68

1. Introduction

Arguments built on symmetry have been used for a very long time in all branches of science. Until the advent of modern quantum mechanics the approach to symmetry was quite simple and largely intuitive, and the arguments were not very different from the way symmetry is used in our everyday life. Shortly after the birth of the Schrödinger equation the use of symmetry was, however, radically changed through the introduction of group theory or rather group representation theory. Long before that group theory had of course been established as a mathematical subject, but it was with the work of people like Weyl, Wigner and many others in the early thirties that group theory was shown to be an important tool of modern physics. Today courses in group theory are part of the curriculum at universities all over the world, and there are numerous textbooks on group theory and its applications in physics and chemistry[1)].

In spite of this one cannot but feel that perhaps the way symmetry is being used today is less than optimal. For one, group theory, or rather group representation theory, is really not simple even for the finite point groups. As a result most textbooks only state the various theorems on matrix representations of groups and refrain from proving them, and the reader is left with a serious gap in the understanding of the whole machinery of representation theory. For continuous groups, in particular the ordinary rotation group, the representation theory is so difficult that most textbooks instead choose to treat spherically symmetric systems via the methods of angular momentum. Angular momentum theory, built on the simple commutation properties of the angular momentum operators, is simple and direct and results are obtained with almost no effort. Unfortunately then, systems that have full rotational symmetry are being treated in a completely different way from systems with less than rotational symmetry.

Even within the narrow field of point group symmetry the conventional theory is less than adequate. It concentrates almost exclusively on the characters of representations, rather than the representations themselves. As a result it can only provide rather loose statements about the vanishing or nonvanishing of matrix elements. Let us illustrate this with an example. Using the character tables found in any textbook on group theory we can only say about the matrix element $\langle E|E|E \rangle$ in D_3 that it is nonvanishing, since the direct product of the representation $E \times E$ contains the representation E. However, the matrix element is in reality 8 quantities, since the E representation is doubly degenerate and group theory should be able to provide a quantitative relationship between these 8 quantitities. In fact it does through the theory of coupling coefficients and the Wigner-Eckart theorem, but the conventional treatment with its emphasis on characters does not provide this information. There seems indeed to be a disparity in the amount of work that is required in understanding the theory and the rather weak results that are conventionally derived from it.

Finally the treatment of systems with halfinteger spin as well as the consequences of time reversal is in most cases not very satisfying. Even the more advanced treatments[1)] then rely on quoting the results of the detailed investigations of Wigner[2)] and Bethe[3)] among others[4)]. As a result, the average reader uses the so called double groups only with a certain feeling of uneasiness, due to a rather limited understanding of the way the theory of these systems has been developed. Because of this coupling coeficients for double groups are not used as often as they could.

In short, it is my opinion, that conventional group representation theory, as applied to problems in atomic and molecular physics, is on the one hand too difficult for the average physicist or chemist to fully master, and on the other that it gives too little information to be really worth the effort.

Is there an alternative to this very unsatisfying situation. Indeed there is, as I shall try to show in this paper. The approach, which I have termed the NONCOM (for non-commuting generator) approach to molecular symmetry, is based simply on finding common eigenfunctions to sets of *non-commuting* operators. For molecular symmetry the operators are the generators of the molecular symmetry group. As will become apparent, I hope, the theory is exceedingly simple and transparent, and treats systems of halfinteger spin (the "double groups") just as easily as integer spin systems. Rather than dealing with all the elements of a symmetry group, we consider only the few generators, and the character table of the conventional theory, is replaced by an eigenvalue table. This table is no more complex than the conventional character table, but gives information corresponding to a complete representation of the symmetry group. Thus, on the basis of the eigenvalue table, coupling coefficients are easily evaluated and they permit the calculation of matrix elements through the Wigner-Eckart theorem, as will also be demonstrated.

The approach described here has been used in a course on molecular physics at Odense University for a number of years, and has also been used in papers on resonance Raman scattering[5] and on local modes of vibration[6]. It was inspired by papers by Golebiewski[7, 8], who used a set of *commuting* symmetry operators to characterize the wavefunctions through their eigenvalues under these operators. Clearly this is completely analogous to the angular momentum theory of full rotational symmetry. Unfortunately Golebiewski's approach is still somewhat complicated. If, however, one takes essentially the same point of view, but considers non-commuting as well as commuting operators, one finds that the theory becomes surprisingly simple and easy to use in practical calculations.

Surprisingly Golebiewski's work seems to have been completely ignored in the literature, which is totally dominated by the conventional approach. After having taught the present approach for several years, and mentioned it in several talks and papers, a colleague drew my attention to the work of Altmann, who in a couple of papers[9, 10], and a book[11], has described an approach based on socalled semi-direct product groups, which, technically, is closely related to the present approach. The point of view is, however, completely different, and Altmann only considers integer spin systems.

2. Molecular Symmetry, Point Groups and Generators

Molecular symmetry is conventionally discussed in terms of point groups. These are sets of symmetry operations, like rotation, reflection and inversion that leave a solid body unchanged. A molecule is, however, not a solid body but a complicated many body system and it is not obvious that a point group is appropriate for characterizing the symmetry properties of such a conglomerate. Indeed, in a strict sense, it is not, as has been emphasized by Hougen[12], Longuet-Higgins[13], Bunker[14] and others. The proper symmetry operations for a molecule are: translation, rotation, inversion, permutation of electrons, and permutation of identical nuclei. The combination of inversion and the set of operations that permute identical nuclei forms the socalled complete nuclear permutation inversion group, CNPI[14]. The molecular Hamiltonian is invariant under this group of symmetry operations, and so any eigenstate could be classified under this group. Unfortunately, for molecules containing just a few identical nuclei, the CNPI group soon becomes very large. As an example, the CNPI group for benzene, C_6H_6, has $6!*6!*2 = 1,036,800$ elements! Most of these symmetry elements do not give rise to any observable physical effects, like level splittings or selection rules. This is because a molecule like benzene is essentially "rigid", that is the molecular configuration only departs slightly from an equilibrium configuration. For a "rigid" molecule only a very small subgroup of the CNPI is relevant for characterizing the states of the molecule. As shown in the works mentioned this subgroup of socalled feasible operations is isomorphic with the point group characteristic of the equilibrium configuration of the molecule. This explains why a point group *can* be used for classifying the symmetry properties of a rigid molecule.

A point group consists of operations that leave a single point invariant. These operations are rotations, inversion and reflections. The various points groups are formed by combining the operators in various ways. The derivation of all the point groups in a systematic way was done by Seitz[15]. Here we shall only list them in a systematic way and discuss the set of symmetry operations that may be used to generate them.

A word about notation. For the points groups we shall generally use the Schonflies notation. In this notation basically the same symbols are used for the point groups as for the symmetry operations. Whenever there is any doubt as to the meaning we shall use a caret (example $\hat{C}_n$) to indicate a symmetry operation.

The simplest point groups are generated by just one element. When this element is a rotation around some axis through an angle of $2\pi/n$ we have a rotation operation of order n, symbolized $\hat{C}_n$. The point group, C_n, consists of this operation and its various powers.

The single generating operation can also be a rotation reflection operation, $\hat{S}_{2n}$, consisting of a rotation through $2\pi/2n$ followed by a reflection in a plane perpendicular to the rotation axis. The group is S_{2n}. The operation S_2 is identical to the inversion operation, $\hat{I}$, and for n odd S_{2n} can also be generated by the two operations $\hat{C}_n$ and $\hat{I}$.

The point groups C_{nh} are generated by the two operations $\hat{C}_n$ and the reflection operation $\hat{\sigma}_h$, a reflection in a plane perpendicular to the rotation axis. Since the product of $\hat{C}_2$ and $\hat{\sigma}_h$ is the inversion operation $\hat{I}$, C_{nh}, for n even can also be generated by the two operations $\hat{C}_n$ and $\hat{I}$.

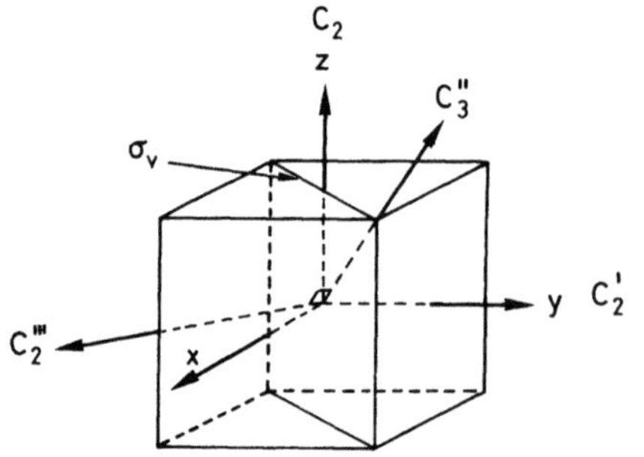

Fig. 1. Location of coordinate axes in cube

Reflection can also be in a plane containing the rotation axis, $\hat{\sigma}_v$. The combination of $\hat{C}_n$ and $\hat{\sigma}_v$ generates the point group C_{nv}.

If instead of a reflection one adds a two-fold rotation $\hat{C}_2'$ perpendicular to the principal rotation axis, one obtains the point group D_n, generated by $\hat{C}_n$ and $\hat{C}_2'$.

When to the above is added a reflection perpendicular to the principal axis, $\hat{\sigma}_h$, one obtains the group D_{nh} generated by $\hat{C}_n$, $\hat{C}_2'$ and $\hat{\sigma}_h$. Again, as for C_{nh}, for n even we can instead generate the point group from the operators $\hat{C}_n$, $\hat{C}_2'$ and $\hat{I}$.

The last of the simple pointgroups is D_{nd}, which conventionally is obtained from D_n by adding reflections in planes that bisect the angles between the $\hat{C}_2'$ axes. It is, however, more convenient to generate D_{nd} from just two operations, $\hat{S}_{2n}$ and $\hat{C}_2'$.

Sofar we have not said anything about the number n. For a free molecule it may indeed be any integer number, including infinity. If the molecule is part of a crystal, however, the translational symmetry of the crystal imposes restrictions on n. From simple geometrical arguments one sees, that in this case only the values 1, 2, 3, 4 and 6 are allowed for n.

As said, a free molecule may have a rotation axis of infinite order. The operation $\hat{C}_\infty$, together with $\hat{\sigma}_v$ generates the point group $C_{\infty v}$ that characterizes hetero-nuclear diatomic molecules. Homonuclear diatomic molecules are characterized by the point group $D_{\infty h}$, generated by $\hat{C}_\infty$, $\hat{C}_2'$ and $\hat{\sigma}_h$, or more conveniently by C_∞, $\hat{C}_2'$ and $\hat{I}$.

What rests are the socalled cubic point groups: T, T_d, T_h, O and O_h. The symmetry operators for these groups are conveniently described with reference to a cube, Fig. 1.

The group T may be generated from D_2 by adding a threefold rotation $\hat{C}_3''$. If the twofold rotations $\hat{C}_2$ and $\hat{C}_2'$ in D_2 are around the z and y axis, respectively, then the threefold rotation is around the xyz diagonal, see Fig. 1.

If to the group T one adds a reflection operation $\hat{\sigma}_v$ in a plane containing the z axis and bisecting the xy angle one obtains the group T_d, the symmetry group of the tetrahedron. Thus, T_d can be generated from $\hat{C}_2$, $\hat{C}_2'$, $\hat{C}_3''$ and σ_v. T_d can, however, also be generated from just two operations, $\hat{S}_4$ and $\hat{C}_3''$, where the fourfold rotation-reflection is around the z-axis of the cube.

Table 1. Point groups and their generators

Point group	Generators	Alternative generators
C_n	$\hat{C}_n$	
S_{2n}	$\hat{S}_{2n}$	$(\hat{C}_n, \hat{I})$ n odd
C_{nh}	$\hat{C}_n, \hat{\sigma}_h$	$(\hat{C}_n, \hat{I})$ n even
C_{nv}	$\hat{C}_n, \hat{\sigma}_v$	
D_n	$\hat{C}_n, \hat{C}_2'$	
D_{nh}	$\hat{C}_n, \hat{C}_2', \hat{\sigma}_h$	$(\hat{C}_n, \hat{C}_2', \hat{I})$ n even
D_{nd}	$\hat{C}_n, \hat{C}_2', \hat{\sigma}_v$	$\hat{S}_{2n}, \hat{C}_2'$
$C_{\infty v}$	$\hat{C}_\infty, \hat{\sigma}_v$	
$D_{\infty h}$	$C_\infty, \hat{C}_2', \hat{\sigma}_h$	$\hat{C}_\infty, \hat{\sigma}_v, \hat{I}$
T	$\hat{C}_2, \hat{C}_2', \hat{C}_3''$	
T_d	$\hat{C}_2, \hat{C}_2', \hat{C}_3'', \hat{\sigma}$	$\hat{S}_4, \hat{C}_3''$
T_h	$\hat{C}_2, \hat{C}_2', \hat{C}_3'', \hat{I}$	
O	$\hat{C}_2, \hat{C}_2', \hat{C}_3'', \hat{C}_2'''$	$\hat{C}_4, \hat{C}_3''$
O_h	$\hat{C}_2, \hat{C}_2', \hat{C}_3'', \hat{C}_2''', \hat{I}$	$\hat{C}_4, \hat{C}_3'', \hat{I}$

$C_i = S_2$; $C_S = C_{1h}$
$C_{1v} = C_{1h}$; $D_1 = C_2$; $D_{1h} = C_{2v}$; $D_{1d} = C_{2h}$

Instead of a reflection operation, one can also add the inversion $\hat{I}$, to the operations of T. Then one obtains the point group T_h.

If to the operations of T is added a twofold rotation $\hat{C}_2'''$ around an axis in the xy plane perpendicular to the C_3'' axis then one obtains the point group O. Again O can also be generated from just two elements, $\hat{C}_4$ and $\hat{C}_3''$.

Finally if to the operations of O is added the inversion one obtains the point group O_h. This is the symmetry group of the octahedron.

In Table 1 we list all the point groups discussed above, with the operators used to generate them. Alternative sets of generators are indicated. Not all the pointgroups listed are unique for all n. It is obvious that C_{1v} is identical to C_{1h}, as are D_1 and C_2, D_{1h} and C_{2v} and D_{1d} and C_{2h}. S_2 is often symbolized C_i; and C_{1h} is C_s.

As said, for the crystallografic point groups n is restricted to the numbers n = 1, 2, 3, 4 and 6. There is thus a total of 32 crystallographic point groups[15].

3. Eigenstates and Commutation Relations

Our goal is to find a set of states that are simultaneous eigenstates for the molecular Hamiltonian and for the set of symmetry operators. Symmetry operators must commute with the Hamiltonian but not necessarily among themselves. A symmetry operator must be unitary, since, obviously, it cannot change the length of a state vector. Suppose we have an eigenstate of the Hamiltonian:

$$H|i\rangle = \varepsilon_i|i\rangle \tag{1}$$

and operate on it with a unitary operator Λ that commutes with the Hamiltonian. Then:

$$H\Lambda|i\rangle = \Lambda H|i\rangle = \varepsilon_i\Lambda|i\rangle \tag{2}$$

showing that the state $\Lambda|i\rangle$ is again an eigenstate of H with the same eigenvalue ε_i. If the eigenvalue ε_i is non-degenerate, then $\Lambda|i\rangle$ can only differ from $|i\rangle$ by a simple phase factor. Thus:

$$\Lambda|i\rangle = \lambda|i\rangle \ , \tag{3}$$

showing that the state $|i\rangle$ is simultaneously an eigenstate of H and of Λ with eigenvalues ε_i and λ, respectively.

If the eigenvalue ε_i is degenerate then the state $\Lambda|i\rangle$ is not necessarily proportional to $|i\rangle$ but may be a linear combination of the set of eigenstates belonging to ε_i. In that case, however, we can make linear combinations of the original states in such a way that the resulting states do become eigenstates also for the Λ operator. This follows directly from the fact that any unitary matrix (the matrix of Λ) can be diagonalised by a unitary transformation. In the following we shall always assume that this transformation has been performed.

For a single symmetry operator we can immediately find the possible eigenvalues. Any symmetry operator Λ is cyclic, that is there exists a number n, such that $\Lambda^n = \hat{E}$, the identity operation. Suppose $|\lambda\rangle$ is an eigenstate of Λ with eigenvalue λ:

$$\Lambda|\lambda\rangle = \lambda|\lambda\rangle \ . \tag{4}$$

Then

$$\Lambda^n|\lambda\rangle = \lambda^n|\lambda\rangle = \hat{E}|\lambda\rangle = 1|\lambda\rangle \ , \tag{5}$$

showing that the eigenvalue λ must be of the form:

$$\lambda = e^{-i\frac{2\pi}{n}\cdot m} \ , m = 0, 1, 2, \ldots\ldots, n-1 \ . \tag{6}$$

An operator of order n thus has precisely n distinct eigenvalues. We are in this section assuming that we are dealing with integer spin systems so that a rotation through 2π equals the identity. Halfinteger spin will be treated in the following section.

For the point groups that are generated by just one symmetry operation we can classify the eigenstates simply according to their eigenvalue under the symmetry operation. For point groups that are generated by more than one symmetry operation, but where the generators commute, we can again choose common eigenstates for the Hamiltonian and for the set of generators. Then the eigenstates are classified simply by the set of eigenvalues under the various generators.

The problem is different when the generators do not commute. As a specific example we can take the case of the point group D_4. The two generators are $\hat{C}_4$ and $\hat{C}_2'$. They do not commute but obey the relation:

$$\hat{C}_4\hat{C}_2' = \hat{C}_2'\hat{C}_4^{-1} \ . \tag{7}$$

This relation follows straightforwardly from geometric inspection.

Suppose now that we have chosen an eigenstate of the Hamiltonian and of the operator $\hat{C}_4$:

$$\hat{C}_4|\lambda_4\rangle = \lambda_4|\lambda_4\rangle \tag{8}$$

We operate on this state with the operator $\hat{C}_2'$ and use the relation, Eq. (7):

$$\hat{C}_4\hat{C}_2'|\lambda_4\rangle = \hat{C}_2'\hat{C}_4^{-1}|\lambda_4\rangle = \lambda_4^{-1}\hat{C}_2'|\lambda_4\rangle \ . \tag{9}$$

Thus, the state $\hat{C}_2'|\lambda_4\rangle$ is again an eigenstate of $\hat{C}_4$ but with eigenvalue λ_4^{-1}. Two possibilities then exist: $\lambda_4 = \pm 1$ so that $\lambda_4 = \lambda_4^{-1}$, or $\lambda_4 \neq \lambda_4^{-1}$.

Let us consider the first case, where $\hat{C}_2'|\lambda_4\rangle$ is an eigenstate of $\hat{C}_4$ with the *same* eigenvalue λ_4. Then again, just as for commuting operators we can choose the states so that they are simultaneous eigenstates of the two generators. Thus, for $\lambda_4 = \lambda_4^{-1}$ a state can be labelled by the two eigenvalues λ_4 and λ_2, so that:

$$\hat{C}_4|\lambda_4, \lambda_2\rangle = \lambda_4|\lambda_4, \lambda_2\rangle; \hat{C}_2'|\lambda_4, \lambda_2\rangle = \lambda_2|\lambda_4, \lambda_2\rangle \ . \tag{10}$$

In the other case, where $\lambda_4 \neq \lambda_4^{-1}$, operation with $\hat{C}_2'$ on the state $|\lambda_4\rangle$ leads to a state with a different eigenvalue under $\hat{C}_4$, namely λ_4^{-1}. The state $\hat{C}_2'|\lambda_4\rangle$ must then, apart from a phase-factor, equal the state $|\lambda_4^{-1}\rangle$. As always in quantum mechanics we can choose phase factors at will, provided it is done consistently. In particular we can choose phases so that:

$$\hat{C}_2'|\lambda_4\rangle = |\lambda_4^{-1}\rangle \ . \tag{11}$$

Then also:

$$\hat{C}_2'|\lambda_4^{-1}\rangle = \hat{C}_2'\hat{C}_2'|\lambda_4\rangle = |\lambda_4\rangle \ . \tag{12}$$

For $\lambda_4 \neq \lambda_4^{-1}$ we see that we cannot give an eigenvalue under the $\hat{C}_2'$ operation. Rather we can give a transformation matrix that describes how the eigenstates under $\hat{C}_4$ behave when operated upon by the operator $\hat{C}_2'$. The form of the matrix is given uniquely from the generalized commutation relation, but the particular matrix elements depend on a phase choice.

For the example treated, the point group D_4, we can now give a complete symmetry classification (for integer spin) in the form of a generalized eigenvalue table, Table 2. In the table we also give the conventional representation symbols. The convention when translated into the eigenvalue language, is that A and B refer to real eigenvalues under $\hat{C}_4$ and E to complex eigenvalues. A stands for eigenvalue 1, B for eigenvalue -1, and indices 1 and 2 refer to the eigenvalues under $\hat{C}_2'$, 1 and -1 respectively.

Precisely the same method as for D_4 can be used to construct generalized eigenvalue tables for all, except the cubic point groups. The generalized commutation relation, Eq. (7), holds for any n

$$\hat{C}_n\hat{C}_2' = \hat{C}_2'\hat{C}_n^{-1} \ , \tag{13}$$

and completely analogous relations are

$$\hat{C}_n\hat{\sigma}_v = \hat{\sigma}_v\hat{C}_n^{-1} \ , \tag{14}$$

$$\hat{S}_{2n}\hat{C}_2' = \hat{C}_2'\hat{S}_{2n}^{-1} \ . \tag{15}$$

Table 2. Eigenvalue table for point group D_4

Representation	Eigenvalue	
Symbol	$\lambda(\hat{C}_4)$	$\lambda(\hat{C}_2')$
A_1	1	1
A_2	1	-1
B_1	-1	1
B_2	-1	-1
E	$-i$ i	$\begin{pmatrix} 0 & 1 \\ 1 & 0 \end{pmatrix}$

The inversion operation commutes with all rotations and reflections and so simply leads to an extra eigenvalue. The operator $\hat{\sigma}_h$ also commutes with both $\hat{C}_n$ and $\hat{C}_2'$ and so again leads to an extra eigenvalue. For all, except the cubic point groups, the construction of the eigenvalue table is then straightforward. Results for all the point groups are listed in Appendix A, in a way similar to Table 2 for D_4.

As said the situation is a little more complex for the cubic point groups. We start with the point group T, which may be generated from the operators $\hat{C}_2$, $\hat{C}_2'$ and $\hat{C}_3''$. The first two operators commute, and the generalized commutation relations with respect to $\hat{C}_3''$ are easily seen to be:

$$\hat{C}_2\hat{C}_3'' = \hat{C}_3''\hat{C}_2' \tag{16}$$

$$\hat{C}_2'\hat{C}_3'' = \hat{C}_3''\hat{C}_2'\hat{C}_2 \, . \tag{17}$$

The first two operators $\hat{C}_2$ and $\hat{C}_2'$ commute and have eigenvalues of ± 1, and so initially we have the four states: $|1,1\rangle$, $|1,-1\rangle$, $|-1,1\rangle$ and $|-1,-1\rangle$. Operating on these with $\hat{C}_3''$ and using Eqs. (16) and (17) we get (now suppressing the $''$ on $\hat{C}_3$ and all caret's for convenience):

$$C_2C_3|1,1\rangle = C_3|1,1\rangle \tag{18a}$$

$$C_2'C_3|1,1\rangle = C_3|1,1\rangle \tag{18b}$$

$$C_2C_3|1,-1\rangle = -C_3|1,-1\rangle \tag{19a}$$

$$C_2'C_3|1,-1\rangle = -C_3|1,-1\rangle \tag{19b}$$

$$C_2C_3|-1,1\rangle = C_3|-1,1\rangle \tag{20a}$$

$$C_2'C_3|-1,1\rangle = -C_3|-1,1\rangle \tag{20b}$$

$$C_2C_3|-1,-1\rangle = -C_3|-1,-1\rangle \tag{21a}$$

$$C_2'C_3|-1,-1\rangle = C_3|-1,-1\rangle \tag{21b}$$

From Eqs. (18) it follows that the state $|1,1\rangle$ can also be taken as an eigenstate of the $\hat{C}_3$ operator. The possible eigenvalues are 1, exp $(\pm i2\pi/3)$, and we obtain the three states: $|1,1,1\rangle$ $|1,1,e^{-i2\pi/3}\rangle$ and $|1,1,e^{+i2\pi/3}\rangle$.

Table 3. Eigenvalue table for point group T

Representation	$\lambda(C_2)$	$\lambda C_2')$	$\lambda(C_3'')$
A	1	1	1
E	1	1	$e^{-i2\pi/3}$
	1	1	$e^{i2\pi/3}$
T	1	−1	$\begin{pmatrix} 0 & 1 & 0 \\ 0 & 0 & 1 \\ 1 & 0 & 0 \end{pmatrix}$
	−1	1	
	−1	−1	

From Eqs. (19)–(21) it follows that the other three states, $|1,-1\rangle$, $|-1,1\rangle$ and $|-1,-1\rangle$ are transformed into each other by the C_3 operation. In particular $\hat{C}_3|1,-1\rangle \sim |-1,-1\rangle$, $\hat{C}_3|-1,1\rangle \sim |1,-1\rangle$ and $\hat{C}_3|-1,-1\rangle \sim |-1,1\rangle$ where the proportionality sign is used to indicate that the states may differ by a phase. Again we can choose phases at will, and an obvious choice would be to choose phases so that the phasefactors become unity. With this choice we obtain the generalized eigenvalue table, Table 3. Two things should be noted. The fact that the two states with complex eigenvalues share the classification E is a consequence of time reversal and is discussed later. Secondly, the transformation matrix has its rows and coloumns arranged such that it gives the matrix of the $\hat{C}_3$ operator. Thus, first row, second coloumn is 1 because: $\langle 1,-1|\hat{C}_3|-1,1\rangle = 1$.

From the group T we obtain the full tetrahedral group T_d by adding the reflection operation σ. The generalized commutation relations are:

$$C_2\sigma = \sigma C_2^{-1} \tag{22}$$

$$C_2'\sigma = \sigma C_2 C_2' \tag{23}$$

$$C_3\sigma = \sigma C_3^{-1} \tag{24}$$

From these it follows that the state $|1,1,1\rangle$ in T is transformed into itself under the reflection operation, and so we obtain the two states $|1,1,1,1\rangle$ and $|1,1,1,-1\rangle$, with the last number specifying the eigenvalue under σ. The two states $|1,1,e^{\pm i2\pi/3}\rangle$ are transformed into each other by the reflection operation. The state $|1,-1\rangle$ in T transforms into itself under the reflection operation, leading again to an extra eigenvalue. The states $|-1,1\rangle$ and $|-1,-1\rangle$ transform into each other under the reflection operation. To find the phase-factors consistently we progress as follows: on operating on the state $|1,-1,-,1\rangle$, where the $-$ refers to the $\hat{C}_3$ operation we get:

$$\begin{aligned} C_3\sigma|1,-1,-,1\rangle &= C_3|1,-1,-,1\rangle = |-1,-1,-,1\rangle \\ &= \sigma_v C_3^{-1}|1,-1,-,1\rangle = \sigma|-1,1,-,1\rangle \ , \end{aligned} \tag{25}$$

showing that:

$$\sigma|-1,1,-,1\rangle = |-1,-1,-,1\rangle \ . \tag{26}$$

Similarly on operating on $|1,-1,-,-1\rangle$

we get

$$\sigma|-1,1,-,-1\rangle = -|-1,-1,-,-1\rangle \ . \tag{27}$$

In this way all phasefactors are fixed and we obtain the eigenvalue table for T_d, Table 4.

The eigenvalue tables for the remaining pointgroups, T_h, O and O_h are constructed in completely analogy with the table for T_d and are also given in Appendix A.

For the construction of the eigenvalue tables it is convenient to use the complex eigenvalues we have employed. Sometimes it is, however, practical to have eigenvalue tables with real elements only. These can be obtained as follows. Let us with $|1\rangle$ and

Table 4. Eigenvalue table for point group T_d

Representation	$\lambda(C_2)$	$\lambda(C_2')$	$\lambda(C_3'')$	$\lambda(\sigma)$
A_1	1	1	1	1
A_2	1	1	1	-1
	1	1	$e^{-i2\pi/3}$	
E				$\begin{pmatrix} 0 & 1 \\ 1 & 0 \end{pmatrix}$
	1	1	$e^{i2\pi/3}$	
	1	-1		-1
T_1	-1	1	$\begin{pmatrix} 0 & 1 & 0 \\ 0 & 0 & 1 \\ 1 & 0 & 0 \end{pmatrix}$	$\begin{pmatrix} 0 & -1 \\ -1 & 0 \end{pmatrix}$
	-1	-1		
	1	-1		1
T_2	-1	1	$\begin{pmatrix} 0 & 1 & 0 \\ 0 & 0 & 1 \\ 1 & 0 & 0 \end{pmatrix}$	$\begin{pmatrix} 0 & 1 \\ 1 & 0 \end{pmatrix}$
	-1	-1		

$|-1\rangle$ ($|2\rangle$ and $|-2\rangle$ for the E_2 representation of sixfold symmetries) denote the states with eigenvalues $e^{-i\omega}$ and $e^{i\omega}$ respectively. Then new states, called $|x\rangle$ and $|y\rangle$ are formed from these by the unitary transformation:

$$\begin{pmatrix} |x\rangle \\ \\ |y\rangle \end{pmatrix} = \begin{pmatrix} -\frac{1}{\sqrt{2}} & \frac{1}{\sqrt{2}} \\ \frac{i}{\sqrt{2}} & \frac{i}{\sqrt{2}} \end{pmatrix} \begin{pmatrix} |1\rangle \\ \\ |-1\rangle \end{pmatrix} \tag{28}$$

The eigenvalue matrix under the principal rotation or rotation-reflection then becomes:

$$\begin{pmatrix} \cos\omega & -\sin\omega \\ \sin\omega & \cos\omega \end{pmatrix}, \tag{29}$$

while under a secondary rotation C_2' or reflection σ_v we get the diagonal eigenvalue matrix:

$$\begin{pmatrix} -1 & 0 \\ 0 & 1 \end{pmatrix}, \tag{30}$$

The two new states obviously correspond to states transforming as x and y respectively, if z is taken to be along the primary rotation axis, and y along the secondary.

4. Half-Integer Spin

To treat systems of half-integer spin we go back to the discussion of the possible eigenvalues for a symmetry operator. If the operator is a rotation operator corresponding to a rotation angle of $2\pi/n$, then an n-fold repetition of the rotation operation corresponds to a total rotation of 2π. For normal functions a rotation through 2π is equal to no operation, that is the identity $\hat{E}$, giving the possible eigenvalues λ as:

$$\lambda = e^{-i\frac{2\pi}{n}\cdot m}, m = 0, 1, 2, \ldots\ldots, n-1 . \tag{6}$$

For halfinteger spin systems, however, rotation through 2π does not equal the identity. Rather, as is well known, half-integer spin states change sign under a 2π rotation and only go into themselves under a 4π rotation. This means, that for halfinteger spin, instead of the condition:

$$\lambda^n = 1 \tag{31}$$

valid for integer spin systems, we have for halfinteger spin:

$$\lambda^n = -1 , \tag{32}$$

leading to the possible eigenvalues:

$$\lambda = e^{-i\frac{2\pi}{n}\cdot m}, \quad m = \pm\frac{1}{2}, \pm\frac{3}{2}, \ldots, \pm\begin{cases}\frac{n}{2} & n \text{ odd}\\ \frac{n-1}{2} & n \text{ even}\end{cases} . \tag{33}$$

This holds for rotation operators. The inversion operation is still a simple operator with eigenvalues ± 1, while reflection, which can be taken as the product of a twofold rotation with inversion, will have the extra eigenvalues $e^{\pm i\pi/2} = \pm i$.

Apart from this change in possible eigenvalues (and some extra complications related to time-reversal symmetry and treated in the following section), the treatment of half-integer spin is exactly similar to the way integer spin was treated. We shall illustrate this with a few examples.

We start again with the pointgroup D_4. The eigenvalues under the $\hat{C}_4$ operation are now: $e^{\pm i\pi/4}$ and $e^{\pm i3\pi/4}$. The commutation relations are as for integer spin and operation with C_2' transforms an eigenstate $|\lambda_4\rangle$ into another eigenstate $|\lambda_4^{-1}\rangle$. Again there is freedom in the choice of phases. Since now, however, we have:

$$\hat{C}_2'\hat{C}_2'|\lambda_4\rangle = -|\lambda_4\rangle , \tag{34}$$

we cannot choose relative phases so that the two states of a pair simply transform into each other with a phasefactor of one. One phasefactor must be -1.

Table 5. Eigenvalue table for D_4, half-integer spin

Representation	$\lambda(\hat{C}_4)$	$\lambda(\hat{C}_2')$
$E_{1/2}$	$e^{-i\pi/4}$ $e^{i\pi/4}$	$\begin{pmatrix} 0 & -1 \\ 1 & 0 \end{pmatrix}$
$E_{3/2}$	$e^{-i3\pi/4}$ $e^{i3\pi/4}$	$\begin{pmatrix} 0 & -1 \\ 1 & 0 \end{pmatrix}$

The eigenvalue table for D_4 is given in Table 5, for half-integer spin. To denote the various types of symmetry we use the same symbols as used by Herzberg. The notation $E_{1/2}$ indicates that the pair of states of angular momentum $J = 1/2$, $M = \pm 1/2$, transforms as that representation, and also suggests that it would be reasonable to choose relative phases for the two components in accordance with the phases for the two components in accordance with the phases of the angular momentum states. For a twofold rotation around the y axis standard angular momentum states give the transformation matrix:

$$\langle JM'|\hat{C}_{2y}|JM\rangle = (-1)^{J-M}\delta_{M',-M} \,. \tag{35}$$

With $J = 1/2, 3/2$ for the $E_{1/2}$, $E_{3/2}$ representations respectively we obtain the transformation matrices of Table 5.

Again the situation is slightly more complicated for the cubic groups. We start with T, generated by $\hat{C}_2$, $\hat{C}_2'$ and $\hat{C}_3''$ with the commutation relations:

$$\hat{C}_2\hat{C}_2' = \hat{C}_2'^{-1}\hat{C}_2 \tag{36}$$

$$\hat{C}_2\hat{C}_3 = \hat{C}_3\hat{C}_2' \tag{37}$$

$$\hat{C}_2'\hat{C}_3 = \hat{C}_3\hat{C}_2'\hat{C}_2 \,. \tag{38}$$

Note, that whereas for integer spin $\hat{C}_2$ and $\hat{C}_2'$ simply commute, for halfinteger spin the correct commutation relation is Eq. (36), which is just a special case of the general relation, Eq. (13). Commutation relations for halfinteger spin are discussed further in Appendix B.

States may first be classified under $\hat{C}_2$ giving the two states $|1/2\rangle$ and $|-1/2\rangle$, of eigenvalues $-i$ and i respectively. Under $\hat{C}_2'$ these two states are transformed into each other, just as for D_4 and with the same transformation matrix. The matrix under $\hat{C}_3$ is written (with m', $m = \pm 1/2$):

$$\langle m'|\hat{C}_3|m\rangle = \begin{pmatrix} C_{11} & C_{12} \\ C_{21} & C_{22} \end{pmatrix} . \tag{39}$$

From:

$$\begin{aligned}\hat{C}_2\hat{C}_3|\tfrac{1}{2}\rangle &= \hat{C}_2[C_{11}|\tfrac{1}{2}\rangle + C_{21}|-\tfrac{1}{2}\rangle] = -iC_{11}|\tfrac{1}{2}\rangle + iC_{21}|-\tfrac{1}{2}\rangle = \hat{C}_3\hat{C}_2'|\tfrac{1}{2}\rangle = \hat{C}_3|-\tfrac{1}{2}\rangle \\ &= C_{12}|\tfrac{1}{2}\rangle + C_{22}|-\tfrac{1}{2}|-\tfrac{1}{2}\rangle\end{aligned} \tag{40}$$

we get:

$$C_{12} = -iC_{11} \,; \quad C_{22} = iC_{21} \,, \tag{41}$$

and from

$$\begin{aligned}\hat{C}_2'\hat{C}_3|\tfrac{1}{2}\rangle &= \hat{C}_2'[C_{11}|\tfrac{1}{2}\rangle + C_{21}|-\tfrac{1}{2}\rangle] = C_{11}|-\tfrac{1}{2}\rangle - C_{21}|\tfrac{1}{2}\rangle = \hat{C}_3\hat{C}_2'\hat{C}_2|\tfrac{1}{2}\rangle \\ &= \hat{C}_3(-i)|-\tfrac{1}{2}\rangle = -iC_{12}|\tfrac{1}{2}\rangle - C_{22}|-\tfrac{1}{2}\rangle\end{aligned} \tag{42}$$

$$C_{11} = -iC_{22} \,; \quad C_{21} = iC_{12} \,, \tag{43}$$

showing that the matrix C has the form:

$$\underline{\underline{C}} = C_{11} \begin{pmatrix} 1 & -i \\ 1 & i \end{pmatrix} \tag{44}$$

Unitarity demands that $|C_{11}| = \frac{1}{\sqrt{2}}$ and so $C_{11} = e^{i\phi}/\sqrt{2}$. Since $\hat{C}_3^3 = -\hat{E}$, for halfinteger spin, we get:

$$\underline{\underline{C}}_3 = e^{i3\phi}/2\sqrt{2} \begin{pmatrix} 1 & -i \\ 1 & i \end{pmatrix}^3 = e^{i3\phi} e^{i\,3\pi/4} \begin{pmatrix} -1 & 0 \\ 0 & -1 \end{pmatrix}, \tag{45}$$

Showing that:

$$3\left(\phi + \frac{\pi}{4}\right) = n \cdot 2\pi , \quad n = 0, 1, 2 \tag{46}$$

We then get the eigenvalue table for T, Table 6. The last two pairs of the table transform into each other under the time-reversal operation which is why they are collected under the common representation $G_{3/2}$.

The last pointgroup to consider is T_d, obtained from T by adding the reflection operation σ. The commutation relations are:

$$C_2\sigma = \sigma C_2^{-1} \tag{47}$$

$$C_2'\sigma = \sigma C_2 C_2' \tag{48}$$

$$C_3\sigma = \sigma C_3^{-1} \tag{49}$$

First consider the two states belonging to the $E_{1/2}$ representation of T. Defining:

$$\sigma|½\rangle = e^{i\alpha}|-½\rangle \; ; \quad \sigma|-½\rangle = e^{i\beta}|½\rangle \tag{50}$$

we get, from $\sigma^2 = -E$, that

$$(\alpha + \beta) = \pm\pi . \tag{51}$$

Table 6. Eigenvalue table for T, halfinteger spin

Representation		$\lambda(\hat{C}_2)$	$\lambda(\hat{C}_2')$	$\lambda(\hat{C}_3)$
$E_{1/2}$	$\frac{1}{2}$ $-\frac{1}{2}$	$-i$ i	$\begin{pmatrix} 0 & -1 \\ 1 & 0 \end{pmatrix}$	$\frac{e^{-i\pi/4}}{\sqrt{2}} \begin{pmatrix} 1 & -i \\ 1 & i \end{pmatrix}$
$G_{3/2}$	$\frac{1}{2}$ $-\frac{1}{2}$	$-i$ i	$\begin{pmatrix} 0 & -1 \\ 1 & 0 \end{pmatrix}$	$\frac{e^{i5\pi/12}}{\sqrt{2}} \begin{pmatrix} 1 & -i \\ 1 & -i \end{pmatrix}$
	$\frac{1}{2}$ $-\frac{1}{2}$	$-i$ i	$\begin{pmatrix} 0 & -1 \\ 1 & 0 \end{pmatrix}$	$\frac{e^{i13\pi/12}}{\sqrt{2}} \begin{pmatrix} 1 & -i \\ 1 & i \end{pmatrix}$

Also:

$$C_2'\sigma|{\textstyle\frac{1}{2}}\rangle = C_2'e^{i\alpha}|-{\textstyle\frac{1}{2}}\rangle = -e^{i\alpha}|{\textstyle\frac{1}{2}}\rangle = \sigma C_2C_2'|{\textstyle\frac{1}{2}}\rangle = ie^{i\beta}|{\textstyle\frac{1}{2}}\rangle \tag{52}$$

showing that:

$$(\beta - \alpha) = \pi/2 \ . \tag{53}$$

From:

$$\begin{aligned} C_3\sigma|{\textstyle\frac{1}{2}}\rangle &= C_3e^{i\alpha}|-{\textstyle\frac{1}{2}}\rangle = e^{i\alpha}e^{-i\pi/4}(-i|{\textstyle\frac{1}{2}}\rangle + i|-{\textstyle\frac{1}{2}}\rangle)/\sqrt{2} \\ &= \sigma C_3^{-1}|{\textstyle\frac{1}{2}}\rangle = e^{i\pi/4}(e^{i\alpha}|-{\textstyle\frac{1}{2}}\rangle + ie^{i\beta}|{\textstyle\frac{1}{2}}\rangle)/\sqrt{2} \ , \end{aligned} \tag{54}$$

one again obtains the condition Eq. (53). Thus, there are two independent solutions for the pair α, β:

$$(\alpha = \pi/4 \ , \quad \beta = 3\pi/4) \ ; \quad (\alpha = -3\pi/4 \ , \quad \beta = -\pi/4) \ , \tag{55}$$

and the representation $E_{1/2}$ of T goes over into the two representations $E_{1/2}$ and $E_{5/2}$ of T_d, using again the nomenclature of Herzberg.

The two pairs of states that collectively in T (and with timeinversion) transform as $G_{3/2}$ can be described as $(|{\textstyle\frac{1}{2}}\rangle_a, |-{\textstyle\frac{1}{2}}\rangle_a)$ and $(|{\textstyle\frac{1}{2}}\rangle_b, |-{\textstyle\frac{1}{2}}\rangle_b)$. Let us investigate whether a pair of states, e.g. $|{\textstyle\frac{1}{2}}\rangle_a$, $|-{\textstyle\frac{1}{2}}\rangle_a$, transforms into itself under the σ operation. Using the same approach as for $E_{1/2}$ we again get the two equations, Eq. (51) and Eq. (53), determining the constants α and β. Operation with C_3 gives:

$$\begin{aligned} C_3\sigma|{\textstyle\frac{1}{2}}\rangle &= C_3e^{i\alpha}|-{\textstyle\frac{1}{2}}\rangle = e^{i\alpha}e^{i\,5\pi/12}(-i|{\textstyle\frac{1}{2}}\rangle + i|-{\textstyle\frac{1}{2}}\rangle)\sqrt{2} = \sigma C_3^{-1}|-{\textstyle\frac{1}{2}}\rangle \\ &= e^{-i\,5\pi/12}(e^{i\alpha}|-{\textstyle\frac{1}{2}}\rangle + ie^{i\beta}|{\textstyle\frac{1}{2}}\rangle)/\sqrt{2} \end{aligned} \tag{56}$$

Comparing the coefficients for the $|-{\textstyle\frac{1}{2}}\rangle$ state we see that we have an inconsistency, since $i \neq e^{-i\,10\pi/12}$. Thus, the reflection operation must connect the two pairs of states that transform separately under T (when the timereversal operation is not taken into account).

To determine the transformation matrix we proceed exactly as before. We now have the four unknown parameters: α, β, γ, δ defined through:

$$\sigma|{\textstyle\frac{1}{2}}\rangle_a = e^{i\alpha}|-{\textstyle\frac{1}{2}}\rangle_b \tag{57}$$

$$\sigma|-{\textstyle\frac{1}{2}}\rangle_b = e^{i\beta}|{\textstyle\frac{1}{2}}\rangle_\alpha \tag{58}$$

$$\sigma|-{\textstyle\frac{1}{2}}\rangle_a = e^{i\gamma}|{\textstyle\frac{1}{2}}\rangle_b \tag{59}$$

$$\sigma|{\textstyle\frac{1}{2}}\rangle_b = e^{i\delta}|-{\textstyle\frac{1}{2}}\rangle_a \ , \tag{60}$$

with

$$e^{i(\alpha+\beta)} = e^{i(\gamma+\delta)} = -1 \ . \tag{61}$$

From:

$$C_2'\sigma|½\rangle_a = C_2'e^{i\alpha}|-½\rangle_b = -e^{i\alpha}|½\rangle_b = \sigma C_2 C_2'|½\rangle_a = \sigma C_2|-½\rangle_a = i\sigma|-½\rangle_a = ie^{i\gamma}|½\rangle_b \quad (62)$$

comes the condition:

$$-e^{i\alpha} = ie^{i\gamma} \quad (63)$$

and similarly on operating on the state $|½\rangle_b$ one obtains:

$$-e^{i\delta} = ie^{i\beta}\ , \quad (64)$$

giving the four equations determining the constants:

$$\alpha + \beta = \gamma + \delta = \pi\ ;$$
$$\gamma - \alpha = \beta - \delta = \pi\ . \quad (65)$$

One of the constants can be chosen at will; one possible solution is:

$$\alpha = \delta = \pi/4\ ; \quad \beta = \gamma = 3\pi/4 \quad (66)$$

5. Consequences of Time Reversal Symmetry

In the absence of an external magnetic field, the Hamiltonian H is a real Operator. Then, the Schrödinger equation for an ordinary wavefunction, will be invariant under the combined operation of time reversal and complex conjugation:

$$H\psi = i\hbar\frac{\partial\psi}{\partial t} \rightarrow H\psi^* = i\hbar\frac{\partial\psi^*}{\partial(-t)}\ . \quad (67)$$

Thus for an ordinary wavefunction time reversal is equivalent to complex conjugation.
For spinors the time reversal operation is not just complex conjugation. To find the effect of the time reversal operator $\hat{T}$ on a general angular momentum state $|j, m\rangle$ we note that $\hat{T}$ anticommutes with any cartesian component of the angular momentum operator:

$$\hat{T}J_z = -J_z\hat{T}\ . \quad (68)$$

This follows immediately from the fact that any coordinate must be unchanged by time reversal, whereas a velocity, of course, must change sign.
Equation (68) shows that when the time reversal operator $\hat{T}$ operates on a state $|j, m\rangle$ it turns it into a state proportional to the state $|j, -m\rangle$:

$$\hat{T}|j, m\rangle = C_m|j, -m\rangle\ , \quad (69)$$

where C_m is some, as yet undetermined, number of magnitude one.

Since $\hat{T}$ is an antiunitary operator:

$$\hat{T}C|a\rangle = C^*\hat{T}|a\rangle \tag{70}$$

where $|a\rangle$ is any state and C any number, we have the relation:

$$J_+\hat{T} = -\hat{T}J_- \ , \tag{71}$$

where $J_\pm = J_x \pm iJ_y$ are the usual step-up/-down operators. Then:

$$\begin{aligned} J_+\hat{T}|m\rangle &= C_m J_+|-m\rangle = C_m\sqrt{j(j+1)+m(-m+1)}\,|-m+1\rangle = -\hat{T}J_-|m\rangle \\ &= -\hat{T}\sqrt{j(j+1)-m(m-1)}\,|m-1\rangle \\ &= -C_{m-1}\sqrt{j(j+1)-m(m-1)}\,|-m+1\rangle \end{aligned} \tag{72}$$

showing that:

$$C_m = -C_{m-1} \ . \tag{73}$$

From this then follows the important consequence that:

$$\hat{T}^2|m\rangle = \hat{T}C_m|-m\rangle = C^*_m C_{-m}|m\rangle = (-1)^{2m}|m\rangle \ , \tag{74}$$

Thus, for jj and so m integer $\hat{T}^2 = \hat{E}$, the identity operation, while for j halfinteger we have $\hat{T}^2 = -\hat{E}$.

From the fact that $\hat{T}$ is antiunitary it follows that for two arbitrary states $|\psi\rangle$ and $|\phi\rangle$:

$$\langle \hat{T}\psi|\hat{T}\phi\rangle = \langle \phi|\psi\rangle \ . \tag{75}$$

Then:

$$\langle \phi|\hat{T}\phi\rangle = \langle \hat{T}^2\phi|\hat{T}\phi\rangle \ , \tag{76}$$

Table 7. Eigenvalue table for T_d, halfinteger spin

Representation	$\lambda(\hat{C}_2)$	$\lambda(\hat{C}'_2)$	$\lambda(\hat{C}_3)$	$\lambda(\sigma)$
$E_{1/2}$	$\begin{matrix} -i \\ i \end{matrix}$	$\begin{pmatrix} 0 & -1 \\ 1 & 0 \end{pmatrix}$	$\frac{e^{-i\pi/4}}{\sqrt{2}}\begin{pmatrix} 1 & -i \\ 1 & i \end{pmatrix}$	$\begin{pmatrix} 0 & e^{i3\pi/4} \\ e^{i\pi/4} & 0 \end{pmatrix}$
$E_{5/2}$	$\begin{matrix} -i \\ i \end{matrix}$	$\begin{pmatrix} 0 & -1 \\ 1 & 0 \end{pmatrix}$	$\frac{e^{-i\pi/4}}{\sqrt{2}}\begin{pmatrix} 1 & -i \\ 1 & i \end{pmatrix}$	$\begin{pmatrix} 0 & e^{-i\pi/4} \\ e^{-i3\pi/4} & 0 \end{pmatrix}$
$G_{3/2}$	$\begin{matrix} -i \\ i \\ -i \\ i \end{matrix}$	$\begin{matrix} \begin{pmatrix} 0 & -1 \\ 1 & 0 \end{pmatrix} \\ \begin{pmatrix} 0 & -1 \\ 1 & 0 \end{pmatrix} \end{matrix}$	$\begin{matrix} \frac{e^{i5\pi/12}}{\sqrt{2}}\begin{pmatrix} 1 & -i \\ 1 & i \end{pmatrix} \\ \frac{e^{i13\pi/12}}{\sqrt{2}}\begin{pmatrix} 1 & -i \\ 1 & i \end{pmatrix} \end{matrix}$	$\begin{pmatrix} 0 & 0 & 0 & e^{i3\pi/4} \\ 0 & 0 & e^{i\pi/4} & 0 \\ 0 & e^{i3\pi/4} & 0 & 0 \\ e^{i\pi/4} & 0 & 0 & 0 \end{pmatrix}$

which for half-integer spin gives:

$$\langle\phi|\hat{T}\phi\rangle = -\langle\phi|\hat{T}\phi\rangle = 0\ , \tag{77}$$

so that any state must be orthogonal to its timereversed "partner".

The timereversal operator $\hat{T}$ commutes with any symmetry operator. Therefore when $\hat{T}$ operates on an eigenstate of a symmetry operator one obtains a state which again is an eigenstate but with the complex conjugate eigenvalue. Even when the eigenvalue is real the timereversal operation leads to a new state for half-integer spin, since as Eq. (77) shows the time-reversed state must be orthogonal to the initial state. This is *not* the case for integer spin, and so in this case $\hat{T}$ has no effect on states with real eigenvalues.

These facts explain why for simple pointgroups, like C_n, S_{2n} and T, states with complex conjugate eigenvalues under a symmetry operation are collected together under one symmetry symbol although there is no spatial symmetry operation that connects them. They also explain why in C_3, for example, there are two different states with the same eigenvalue, -1, under the spatial symmetry operation. These facts are, of course, just specific examples illustrating Kramers theorem[16].

6. Matrix Elements and Coupling Coefficients

Although we have mainly talked about states it is clear that operators may just as well be classified by their behaviour under the various symmetry operations. Using the eigenvalue tables it is then straight forward to derive quantitative relations between matrix elements.

Consider an arbitrary matrix element $\langle a|\hat{c}|b\rangle$. Since a matrix element is a number it must be unchanged by any symmetry operation. If both states $|a\rangle$ and $|b\rangle$ and the operator $\hat{c}$ have eigenvalues λ_a, λ_b, λ_c respectively under some symmetry operator, then the matrix element can only be non zero if $\lambda_b \cdot \lambda_c = \lambda_a$. If the states or the operator are not eigenstates of the symmetry operator but instead are transformed into other states/operators then quantitative relationships between the matrix elements are obtained.

As an example let us consider a matrix element of the type $\langle E|E|E\rangle$ of d_3. Since E has two components, $|E_1\rangle$ and $|E_{-1}\rangle$, we have a total of 8 matrix elements. Under the $\hat{C}_3$ operation the two components have eigenvalues of $e^{\mp i2\pi/3}$. Thus, using the rule that $\lambda_b \cdot \lambda_c = \lambda_a$ we see that the only nonvanishing matrix elements are $\langle E_{-1}|E_1|E_1\rangle$ and $\langle E_1|E_{-1}|E_{-1}\rangle$. Operating with $\hat{C}_2'$ on $\langle E_{-1}|E_1|E_1\rangle$ gives $\langle E_1|E_{-1}|E_{-1}\rangle$ and so the two matrix elements are identical. In the same way operation with $\hat{C}_3$ shows that the matrix elements $\langle A_1|E_1|E_{-1}\rangle$, $\langle A_1|E_{-1}|E_1\rangle$, $\langle A_2|E_1|E_{-1}\rangle$ and $\langle A_2|E_{-1}|E_1\rangle$ may be nonvanishing, while the $\hat{C}_2'$ operation leads to the relations: $\langle A_1|E_1|_{-1}\rangle = \langle A_1|E_{-1}|E_1\rangle$ and $\langle A_2|E_1|E_{-1}\rangle = -A_2|E_{-1}|E_1\rangle$. Similar relations are obtained for all the other pointgroups in the same way, and as it is seen, with almost no effort.

The same results may also be formulated in terms of coupling coefficients. From the eigenvalue tables we can immediately see how a product of two states transforms under the various symmetry operations. Let us consider the product of two E states in the point group D_3. We have the four product states: $|E_1E_1\rangle$, $|E_1E_{-1}\rangle$, $|E_{-1}E_1\rangle$ and $|E_{-1}E_{-1}\rangle$.

Under the $\hat{C}_3$ operation these states are again eigenstates with eigenvalues equal to the product of the individual eigenvalues, whereas operation with $\hat{C}_2'$ connects $|E_1E_1\rangle$ with $|E_{-1}E_{-1}\rangle$ and $|E_1E_{-1}\rangle$ with $|E_{-1}E_1\rangle$. Forming the symmetric and antisymmetric products we see that $\frac{1}{\sqrt{2}}\{|E_1E_{-1}\rangle \pm |E_{-1}E_1\rangle\}$ transforms as $|A_1\rangle$ and $|A_2\rangle$ respectively, while the pair: $|E_{-1}E_{-1}\rangle$ and $|E_1E_1\rangle$ transforms as the $|E_1\rangle$, $|E_{-1}\rangle$ pair of the E representation. Thus we can write:

$$\frac{1}{\sqrt{2}}\{|E_1E_{-1}\rangle + |E_{-1}E_1\rangle\} = e^{i\alpha}|A_1\rangle \tag{78 a}$$

$$\frac{1}{\sqrt{2}}\{|E_1E_{-1}\rangle - |E_{-1}E_1\rangle\} = e^{i\beta}|A_2\rangle \tag{78 b}$$

$$\begin{pmatrix}|E_{-1}E_{-1}\rangle\\ |E_1\ \ E_1\rangle\end{pmatrix} = e^{i\gamma}\begin{pmatrix}|E_1\rangle\\ |E_{-1}\rangle\end{pmatrix}, \tag{78 c}$$

where $e^{i\alpha}$, $e^{i\beta}$ and $e^{i\gamma}$ arc arbitrary phase-factors, and $1/\sqrt{2}$ is used for purpose of normalization. For convenience we may choose the phasefactors equal to one. Then from Eqs. (78 a, b) we have

$$|E_1E_{-1}\rangle = \frac{1}{\sqrt{2}}(|A_1\rangle + |A_2\rangle) \tag{79 a}$$

$$|E_{-1}E_1\rangle = \frac{1}{\sqrt{2}}(|A_1\rangle - |A_2\rangle) \tag{79 b}$$

and from Eq. (78 c)

$$|E_{-1}E_{-1}\rangle = |E_1\rangle \tag{79 c}$$

$$|E_1E_1\rangle = |E_{-1}\rangle\,. \tag{79 d}$$

The nonvanishing coupling coefficients are then:

$$\langle A_1|E_1E_{-1}\rangle = \frac{1}{\sqrt{2}} \tag{80 a}$$

$$\langle A_1|E_{-1}E_1\rangle = \frac{1}{\sqrt{2}} \tag{80 b}$$

$$\langle A_2|E_1E_{-1}\rangle = \frac{1}{\sqrt{2}} \tag{80 c}$$

$$\langle A_2|E_{-1}E_1\rangle = -\frac{1}{\sqrt{2}} \tag{80 d}$$

$$\langle E_1|E_{-1}E_{-1}\rangle = 1 \tag{80e}$$

$$\langle E_{-1}|E_1E_1\rangle = 1\,, \tag{80f}$$

giving again the relations between matrix elements derived above. In the coupling coefficients the order of the three factors is most important. One can choose phases in such a way that the coupling coefficients have simple symmetry relations upon permutation of the factors. This has been discussed by numerous authors [17, 18]. For the derivation of relations between matrix elements, which is the all important purpose of using coupling coefficients, the direct method we have discussed above is, however, much easier and avoids the problem of phases and coupling coefficients completely.

Acknowledgment. Kjeld Rasmussen, Kemisk Laboratorium A, Danmarks Tekniske Højskole, is thanked for drawing my attention to the work of Altmann.

Appendix A. Eigenvalue Tables

C_2	$\lambda(\hat{C}_2)$
A	1
B	-1
$E_{1/2}$	$-i$
	i

C_s	$\lambda(\hat{\sigma}_n)$
A′	1
A″	-1
$E_{1/2}$	$-i$
	i

C_i	$\lambda(\hat{I})$
Ag	1
A_u	-1
$E_{1/2g}$	1
	1
$E_{1/2u}$	-1
	-1

C_3	$\lambda(\hat{C}_3)$
A	1
E	$e^{-i2\pi/3}$
	$e^{i2\pi/3}$
$E_{1/2}$	$e^{-i\pi/3}$
	$e^{i\pi/3}$
$E_{3/2}$	-1
	-1

S_4 / C_4	$\lambda(\hat{S}_4)$ / $\lambda(\hat{C}_4)$
A	1
B	-1
E	$-i$
	i
$E_{1/2}$	$e^{-i\pi/4}$
	$e^{i\pi/4}$
$E_{3/2}$	$e^{-i3\pi/4}$
	$e^{i3\pi/4}$

C_6	$\lambda(\hat{C}_6)$
A	1
B	-1
E_1	$e^{-i2\pi/6}$
	$e^{i2\pi/6}$
E_2	$e^{-i2\pi/3}$
	$e^{i2\pi/3}$
$E_{1/2}$	$e^{-i\pi/6}$
	$e^{i\pi/6}$
$E_{3/2}$	$e^{-i\pi/2}$
	$e^{i\pi/2}$
$E_{5/2}$	$e^{-i5\pi/6}$
	$e^{i5\pi/6}$

$C_{2h} = C_2 * \hat{I}$; $C_{4h} = C_4 * \hat{I}$; $C_{6h} = C_6 * \hat{I}$

C_{3h}	$\lambda(\hat{C}_3)$	$\lambda(\hat{\sigma}_n)$
A'	1	1
A''	1	-1
E'	$e^{-i\,2\pi/3}$	1
	$e^{i\,2\pi/3}$	1
E''	$e^{-i\,2\pi/3}$	-1
	$e^{i\,2\pi/3}$	-1
$E_{1/2}$	$e^{-i\,\pi/3}$	$-i$
	$e^{i\,\pi/3}$	i
$E_{5/2}$	$e^{i\,\pi/3}$	$-i$
	$e^{-i\,\pi/3}$	i
$E_{3/2}$	$e^{-i\pi}$	i
	$e^{i\pi}$	$-i$

D_2	$\lambda(\hat{C}_2)$	$\lambda(\hat{C}_2')$
C_{2v}	$\lambda(\hat{C}_2)$	$\lambda(\hat{\sigma}_v)$
A_1	1	1
A_2	1	-1
B_1	-1	1
B_2	-1	-1
$E_{1/2}$	$\begin{cases}-i\\ i\end{cases}$	$\begin{pmatrix}0 & -1\\ 1 & 0\end{pmatrix}$

D_3	$\lambda(\hat{C}_3)$	$\lambda(\hat{C}_2')$
C_{3v}	$\lambda(\hat{C}_3)$	$\lambda(\hat{\sigma}_v)$
A_3	1	1
A_2	1	-1
E	$e^{-i\,2\pi/3}$	$\begin{pmatrix}0 & 1\\ 1 & 0\end{pmatrix}$
	$e^{i\,2\pi/3}$	
$E_{1/2}$	$e^{-i\,\pi/3}$	$\begin{pmatrix}0 & -1\\ 1 & 0\end{pmatrix}$
	$e^{i\,\pi/3}$	
$E_{3/2}$	$e^{-i\pi}$	$\begin{pmatrix}0 & -1\\ 1 & 0\end{pmatrix}$
	$e^{i\pi}$	

D_{2d}	$\lambda(\hat{S}_4)$	$\lambda(\hat{C}_2')$
D_4	$\lambda(C_4)$	$\lambda(C_2')$
C_{4v}	$\lambda(\hat{C}_4)$	$\lambda(\hat{\sigma}_v)$
A_1	1	1
A_2	1	-1
B_1	-1	1
B_2	-1	-1
E	$-i$	$\begin{pmatrix}0 & 1\\ 1 & 0\end{pmatrix}$
	i	
$E_{1/2}$	$e^{-i\,\pi/4}$	$\begin{pmatrix}0 & -1\\ 1 & 0\end{pmatrix}$
	$e^{i\,\pi/4}$	
$E_{3/2}$	$e^{-i\,3\pi/4}$	$\begin{pmatrix}0 & -1\\ 1 & 0\end{pmatrix}$
	$e^{i\,3\pi/4}$	

D_6 C_{6v}	$\lambda(\hat{C}_6)$ $\lambda(\hat{C}_6)$	$\lambda(\hat{C}_2')$ $\lambda(\hat{\sigma}_v)$
A_1	1	1
A_2	1	-1
B_1	-1	1
B_2	-1	-1
E_1	$e^{-i2\pi/6}$ $e^{i2\pi/6}$	$\begin{pmatrix}0&1\\1&0\end{pmatrix}$
E_2	$e^{-i2\pi/3}$ $e^{i2\pi/3}$	$\begin{pmatrix}0&1\\1&0\end{pmatrix}$
$E_{1/2}$	$e^{-i\pi/6}$ $e^{i\pi/6}$	$\begin{pmatrix}0&-1\\1&0\end{pmatrix}$
$E_{3/2}$	$e^{-i\pi/2}$ $e^{i\pi/2}$	$\begin{pmatrix}0&-1\\1&0\end{pmatrix}$
$E_{5/2}$	$e^{-i5\pi/6}$ $e^{i5\pi/6}$	$\begin{pmatrix}0&-1\\1&0\end{pmatrix}$

$D_{2h} = D_2 * \hat{I}$; $D_{4h} = D_4 * \hat{I}$;
$D_{6h} = D_6 * \hat{I}$.

D_{3h}	$\lambda(\hat{C}_3)$	$\lambda(\hat{C}_2')$	$\lambda(\hat{\sigma}_h)$
A_1'	1	1	1
A_1''	1	1	-1
A_2''	1	1	-1
E'	$e^{-i2\pi/3}$ $e^{i2\pi/3}$	$\begin{pmatrix}0&1\\1&0\end{pmatrix}$	1 1
E''	$e^{-i2\pi/3}$ $e^{i2\pi/3}$	$\begin{pmatrix}0&1\\1&0\end{pmatrix}$	-1 -1
$E_{1/2}$	$e^{-i\pi/3}$ $e^{-i\pi/3}$	$\begin{pmatrix}0&-1\\1&0\end{pmatrix}$	$-i$ i
$E_{5/2}$	$e^{i\pi/3}$ $e^{-i\pi/3}$	$\begin{pmatrix}0&-1\\1&0\end{pmatrix}$	$-i$ i
$E_{3/2}$	$e^{-i\pi}$ $e^{i\pi}$	$\begin{pmatrix}0&-1\\1&0\end{pmatrix}$	i $-i$

$S_6 = C_3 * \hat{I}$; $D_{3d} = D_3 * \hat{I}$.

D_∞ $C_{\infty v}$	$\lambda(\hat{C}_\phi)$ $\lambda(\hat{C}_\phi)$	$\lambda(\hat{\sigma}_v)$ $\lambda(\hat{\sigma}_v)$
Σ^+	1	1
Σ^-	1	-1
π	$e^{-i\phi}$ $e^{i\phi}$	$\begin{pmatrix}0&1\\1&0\end{pmatrix}$
Δ	$e^{-i2\phi}$ $e^{i2\phi}$	$\begin{pmatrix}0&1\\1&0\end{pmatrix}$
$\vdots$		
$E_{1/2}$	$e^{-i\phi\cdot 1/2}$ $e^{i\phi\cdot 1/2}$	$\begin{pmatrix}0&-1\\1&0\end{pmatrix}$
$E_{3/2}$	$e^{-i\phi\cdot 3/2}$ $e^{i\phi\cdot 3/2}$	$\begin{pmatrix}0&-1\\1&0\end{pmatrix}$

$D_{\infty h} = c_{\infty v} * \hat{I}$.

T	$\lambda(\hat{c}_2)$	$\lambda(C'_2)$	$\lambda(C''_3)$
A	1	1	1
E	1	1	$e^{-i2\pi/3}$
	1	1	$e^{i2\pi/3}$
T	1	−1	$\begin{pmatrix} 0 & 1 & 0 \\ 0 & 0 & 1 \\ 1 & 0 & 0 \end{pmatrix}$
	−1	1	
	−1	−1	
$E_{1/2}$	$\begin{cases} -i \\ i \end{cases}$	$\begin{pmatrix} 0 & -1 \\ 1 & 0 \end{pmatrix}$	$\frac{e^{-i\pi/4}}{\sqrt{2}} \begin{pmatrix} 1 & -i \\ 1 & i \end{pmatrix}$
$G^{3/2}$	$\begin{cases} -i \\ i \end{cases}$	$\begin{pmatrix} 0 & -1 \\ 1 & 0 \end{pmatrix}$	$\frac{e^{i5\pi/12}}{\sqrt{2}} \begin{pmatrix} 1 & -i \\ 1 & i \end{pmatrix}$
	$\begin{cases} -i \\ i \end{cases}$	$\begin{pmatrix} 0 & -1 \\ 1 & 0 \end{pmatrix}$	$\frac{e^{i13\pi/12}}{\sqrt{2}} \begin{pmatrix} 1 & -i \\ 1 & i \end{pmatrix}$

O T_d	$\lambda(\hat{C}_2)$ $\lambda(\hat{C}_2)$	$\lambda(\hat{C}'_2)$ $\lambda(\hat{C}'_2)$	$\lambda(\hat{C}'_3)$ $\lambda(\hat{C}''_3)$	$\lambda(\hat{C}'''_2)$ $\lambda(\sigma)$
A_1	1	1	1	1
A_2	1	1	1	−1
E	1	1	$e^{-i2\pi/3}$	$\begin{pmatrix} 0 & 1 \\ 1 & 0 \end{pmatrix}$
	1	1	$e^{i2\pi/3}$	
T_1	1	−1	$\begin{pmatrix} 0 & 1 & 0 \\ 0 & 0 & 1 \\ 1 & 0 & 0 \end{pmatrix}$	−1
	−1	1		$\begin{pmatrix} 0 & -1 \\ -1 & 0 \end{pmatrix}$
	−1	−1		
T_2	1	−1	$\begin{pmatrix} 0 & 1 & 0 \\ 0 & 0 & 1 \\ 1 & 0 & 0 \end{pmatrix}$	1
	−1	1		$\begin{pmatrix} 0 & 1 \\ 1 & 0 \end{pmatrix}$
	−1	1		
$E_{1/2}$	$\begin{cases} -i \\ i \end{cases}$	$\begin{pmatrix} 0 & -1 \\ 1 & 0 \end{pmatrix}$	$\frac{e^{-i\pi/4}}{\sqrt{2}} \begin{pmatrix} 1 & -i \\ 1 & i \end{pmatrix}$	$\begin{pmatrix} 0 & e^{i3\pi/4} \\ e^{i\pi/4} & 0 \end{pmatrix}$
$E_{5/2}$	$\begin{cases} -i \\ i \end{cases}$	$\begin{pmatrix} 0 & -1 \\ 1 & 0 \end{pmatrix}$	$\frac{e^{-i\pi/4}}{\sqrt{2}} \begin{pmatrix} 1 & -i \\ 1 & i \end{pmatrix}$	$\begin{pmatrix} 0 & e^{-i\pi/4} \\ e^{-i3\pi/4} & 0 \end{pmatrix}$
$G_{3/2}$	$\begin{cases} -i \\ i \end{cases}$	$\begin{pmatrix} 0 & -1 \\ 1 & 0 \end{pmatrix}$	$\frac{e^{i5\pi/12}}{\sqrt{2}} \begin{pmatrix} 1 & -i \\ 1 & i \end{pmatrix}$	$\begin{pmatrix} 0 & 0 & 0 & e^{i3\pi/4} \\ 0 & 0 & e^{i\pi/4} & 0 \\ 0 & e^{i3\pi/4} & 0 & 0 \\ e^{i\pi/4} & 0 & 0 & 0 \end{pmatrix}$
	$\begin{cases} -i \\ i \end{cases}$	$\begin{pmatrix} 0 & -1 \\ 1 & 0 \end{pmatrix}$	$\frac{e^{i13\pi/12}}{\sqrt{2}} \begin{pmatrix} 1 & -i \\ 1 & i \end{pmatrix}$	

$T_h = T * \hat{I}$; $O_h = 0 * \hat{I}$.

Appendix B. Commutation Relations and Half-Integer Spin

Commutation relations for symmetry operators operating on ordinary functions can be found by direct inspection. This is not the case when we are dealing with half-integer spin functions, since these, as is well known, change sign under a rotation through 2π. Then, the commutation relations must be found by direct calculation. Since the behaviour under rotation through 2π is common to all half-integer spin functions the commutation relations can be found from their action on the simplest functions, $s = \frac{1}{2}$, $m_s = \pm\frac{1}{2}$.

The rotation operator for rotation through the angle χ around a unit axis $\vec{\varepsilon}$ is given by:

$$\hat{R}(\bar{\varepsilon}, \chi) = e^{-i/\hbar\, \chi\vec{\varepsilon} \cdot \vec{J}}\,, \tag{B1}$$

where $\bar{J}$ is the angular momentum operator.

Expanding the dot-product:

$$(\vec{\varepsilon} \cdot \vec{J} = \varepsilon_z J_z + \tfrac{1}{2}(\varepsilon_+ J_- + \varepsilon_- J_+) \tag{B2}$$

with

$$\varepsilon_\pm = \varepsilon_x \pm i\,\varepsilon_y;\; J_\pm = J_x \pm i\,J_y\,, \tag{B3}$$

and using the step-up (step-down) properties of $J_+(J_-)$ gives the matrix $\langle m'|\varepsilon \cdot J/\hbar|m\rangle$ for $s = \frac{1}{2}$:

m' \ m	$\frac{1}{2}$	$\frac{1}{2}$
$\frac{1}{2}$	$\varepsilon_z/2$	$\varepsilon_-/2$
$-\frac{1}{2}$	$\varepsilon_+/2$	$-\varepsilon_z/2$

(B4)

From this the matrix of the rotation operator is readily calculated:

$$\langle m'|e^{-i/\hbar\, \chi\vec{\varepsilon}\cdot\vec{J}}|m\rangle = \langle m'|\sum_{k=0}^{\infty}\left(-\frac{i}{\hbar}\chi\vec{\varepsilon}\cdot\vec{J}\right)^k \frac{1}{k!}\,|m\rangle$$

$$= \sum_{k=0}^{\infty}(-i\chi/2)^k\,\frac{1}{k!}\begin{pmatrix}\varepsilon_z & \varepsilon_-\\ \varepsilon_+ & -\varepsilon_z\end{pmatrix}^k$$

$$= \left\{1 + (-i\chi/2)^2\cdot\frac{1}{2!} + \ldots.\right\}\begin{pmatrix}1 & 0\\ 0 & 1\end{pmatrix} + \left\{-i\chi(2) + (-i\chi/2)^3\frac{1}{3!}\right\}\begin{pmatrix}\varepsilon_z & \varepsilon_-\\ \varepsilon_+ & -\varepsilon_z\end{pmatrix}$$

$$= \cos\frac{\chi}{2}\begin{pmatrix}1 & 0\\ 0 & 1\end{pmatrix} - i\sin(\chi/2)\begin{pmatrix}\varepsilon_z & \varepsilon_-\\ \varepsilon_+ & -\varepsilon_z\end{pmatrix}. \tag{B5}$$

The various commutation relations can be directly calculated from this expression. As an example we consider the two operators $\hat{C}_{nz}$ and $\hat{C}_{2y}$. We get the matrices:

$$\hat{C}_{nz} : \begin{pmatrix} e^{-i\pi/n} & 0 \\ 0 & e^{i\pi/n} \end{pmatrix}$$

$$\hat{C}_{nz}^{-1} : \begin{pmatrix} e^{i\pi/n} & 0 \\ 0 & e^{-i\pi/n} \end{pmatrix} \tag{B 6}$$

$$\hat{C}_{2y} : \begin{pmatrix} 0 & -1 \\ 1 & 0 \end{pmatrix}$$

from which follows by matrix multiplication the commutation relation:

$$\hat{C}_{nz}\hat{C}_{2y} = C_{2y}\hat{C}_{nz}^{-1} \tag{B 7}$$

All other commutation relations used in the text have been evaluated in this way.

Appendix C. Symmetry Properties of Angular Momentum States

In many cases of practical interest we need to know how angular momentum states $|J, M\rangle$ behave under the symmetry operations that characterize the point groups. Examples are atoms in molecules, molecular orbital theory, ligand field theory and the theory of spin-forbidden transitions, to name a few. In this appendix we set forth the necessary formula needed to handle these problems. Some of the expressions may be found in any standard textbook on angular momentum theory, but others, in particular the relations between the Euler angles and the orientation- and rotation angles for arbitrary rotation operators are difficult to find. Also conventions vary widely and errors abound. Therefore the needed expressions are given here in what is thought to be a convenient form.

From the theory of the rotation group we know that any rotation may be specified by the three Euler angels α, β, γ. The operator that describes the *active* rotation of a state through the Euler angles is:

$$R(\alpha, \beta\, \gamma) = e^{-i\alpha J_z/\hbar} e^{-\beta J_y/\hbar} e^{-\gamma j_z/\hbar} \,, \tag{C 1}$$

giving the rotation matrix:

$$\begin{aligned} D^J_{M'M} &= \langle JM'|e^{-i\alpha J_z/\hbar} e^{-i\beta J_y/\hbar} e^{-i\gamma J_z/\hbar}|JM\rangle \\ &= e^{-i\alpha M'} e^{-i\gamma M} \langle JM'|e^{-i\beta J_y/\hbar}|JM\rangle = e^{-i\alpha M'} e^{-i\gamma M} d^J_{M'M}(\beta). \end{aligned} \tag{C 2}$$

The rotation matrix may be calculated recursively by angular momentum coupling from the rotation matrix for $J = \frac{1}{2}$, or from the closed form first derived by Wigner:

$$d^J_{M'M}(\beta) = \sum_t (-1)^t \frac{\sqrt{(J+M')!(J-M')!(J+M)!(J-M)!}}{(J+M'-t)!(J-M-t)!t!(t-M'+M)!} \left(\cos\frac{\beta}{2}\right)^{2J+M'-M-2t} \left(\sin\frac{\beta}{2}\right)^{2t-M'+M} \tag{C3}$$

where the summation over t is restricted to those values for which the argument of any factorial is nonnegative.

In dealing with finite rotations one normally does not specify the three Euler angles (α, β, γ), but rather the rotation angle χ and the two polar angles θ and ϕ defining the orientation of the rotation axis. The connection between the two sets of angles is:

$$\phi = \frac{\alpha - \gamma}{2} + \frac{\pi}{2} \tag{C4a}$$

$$\cos\frac{\chi}{2} = \cos\frac{\beta}{2}\cos\frac{\alpha+\gamma}{2} \tag{C4b}$$

$$\sin\frac{\chi}{2}\sin\theta = \sin\frac{\beta}{2} \tag{C4c}$$

For the non-cubic point groups the symmetry operators are:

1) Rotations around the principal axis which we take to be the z axis
2) rotations around the secondary axis of π, we take this axis as the y axis
3) inversion
4) σ_v, operations which we replace with the product of C_2' and I
5) σ_h operations which similarly equals the product of C_2 and I.

For all these operations the only non-tivial matrixelement of the rotation operator is

$$d^J_{M'M}(\pi) = (-1)^{J-M}\delta_{M',-M}\,. \tag{C5}$$

For the cubic groups the only problem lies with the C_3'' operator. For this operator $\chi = 2\pi/3$, $\sin\theta = \sqrt{\frac{2}{3}}$ and $\phi = \frac{\pi}{4}$ giving the Euler angles: $\alpha = 0$; $\beta = \frac{\pi}{2}$; $\gamma = \frac{\pi}{2}$.

Similarly a rotation of $2\pi/4$ along any axis lying in the $x-y$ plane also gives $\beta = \frac{\pi}{2}$.

Thus, only $d^J_{M'M}\left(\frac{\pi}{2}\right)$ need be calculated explicitly in the most general case. This is done very simply on a computer.

From the rotation matrix it is now trivial to form those linear combinations of the $|J, M\rangle$ states that are eigenstates under the site-symmetry operations.

In a formal way this may be done through projection operators. The projection operator corresponding to the m'th eigenvalue $e^{-i2\pi m/n}$ for a symmetry operator of order n is simply:

$$P_m = \prod_{k\neq m}(\Lambda - \lambda_k) = \prod_{k\neq m}(\Lambda - e^{-i2\pi k/n})\,, \tag{C6}$$

which follows from the fact that it "kills" off all other components.

In practice it is, however, often easier to find the desired function by inspection.

References

1. Tinkham, M.: Group Theory and Quantum Mechanics, McGraw-Hill, New York 1964
2. Wigner, E. P.: Göttingen Nachrichten, Math.-Physik. Kl. 546 (1932)
3. Bethe, H.: Ann. der Physik *3,* 133 (1929)
4. Opechowski, W.: Physica *7,* 552 (1940)
5. Mortensen, O. Sonnich, Hassing, S.: in Advances in Infrared and Raman spectroscopy (ed. Clark, R. J. H., Hester, R. E.) Heyden, London 1979, p. 1
6. Henry, B R., Tarr, A. W., Mortensen, O. Sonnich, Murphy W. F., Compton, D. A. C.: J. Chem. Phys. *79,* 2583 (1983)
7. Golebiewski, A.: Acta Phys. Pol. *A37,* 871 (1970)
8. Golebiewski, A.: Mol. Phys. *20,* 481 (1971)
9. Altmann, S.: Phil. Trans. Roy. Soc. (London) *A255,* 216 (1963)
10. Altmann, S.: Rev. Mod. Phys. *35,* 641 (1963)
11. Altmann, S.: Induced Representations in Crystals and Molecules, Academic Press, London 1977
12. Hougen, J. T.: J. Chem. Phys. *37,* 1433 (1962)
13. Longuet-Higgins, H. C.: Mol. Phys. *6,* 445 (1963)
14. Bunker, P. R.: in Vibrational Spectra (ed. Durig, J. R.) Elsevier, Amsterdam 1975, p. 1
15. Seitz, F.: Z. Krist. *88,* 433 (1934); *90,* 289 (1935); *91,* 336 (1935); *94,* 100 (1936)
16. Kramers, H. A.: Proc. Amsterdam Acad. *33,* 959 (1930)
17. Griffith, J. S.: The Irreducible Tensor Method for Molecular Symmetry Groups, Prentice-Hall, New York 1962
18. Harnung, S. E.: Mol. Phys. *26,* 473 (1973)

Theoretical Models of Cluster Bonding

D. Michael P. Mingos and Roy L. Johnston

Inorganic Chemistry Laboratory, University of Oxford, South Parks Road, Oxford OX1 3QR, United Kingdom

The theoretical models which have been used to describe the bonding in cluster compounds of the main group and transition metal elements are reviewed. The historical development of these models is outlined and special emphasis is placed on those studies which have led to the elucidation of structure-electron count correlations. Theoretical treatments of cluster bonding are based on localised, delocalised (molecular orbital) or free electron methods derived from the solution of the Schrödinger equation for a particle on a sphere. A detailed analysis of the Tensor Surface Harmonic method, as an example of a free electron model, is presented. Group theoretical consequences of the model are also presented.

Structure Bonding 68

A. Introduction

The past four decades have seen the development of a major area of chemistry based on polyhedral clusters of the main group and transition metal elements[1]. In the main group area, the foundations were laid with the characterisation of the polyhedral boranes and carboranes[2], where research was stimulated, in part, by the need for alternative rocket fuels[3]. This has been followed by the synthesis of numerous (and at times isoelectronic and isostructural) bare metal cluster cations and anions of bismuth, tin and other B-Group metals[4]. In addition organic chemists have synthesised many polyhedral $(CH)_n$ molecules as part of the development of the alicyclic chemistry of carbon[5].

In the transition metal field, the unusual catalytic properties of some high nuclearity cluster compounds of rhodium[6] and the belief that transition metal carbonyl clusters[7] could serve as effective models for the chemisorbed state of metal surfaces involved in catalytic processes[8] has led to the study of clusters with up to 55 metal atoms[9]. Cluster compounds of the platinum[10] and coinage metals[11] are generally found to be most stable with ligands (e.g. carbon monoxide, isocyanides and phosphines) which are capable of stabilising low oxidation states. For the earlier transition metals, π-donor ligands such as Cl, S and OR (R = alkyl or aryl group) stabilise clusters with somewhat higher oxidation states[12]. These clusters have been shown to be of considerable structural interest, forming extended arrays in the solid state[13], some of which, such as the Chevrel Phases $M_xMo_6S_8$[14], have been shown to possess unusual magnetic and superconducting properties[15]. The study of molecular sulphido clusters of iron and molybdenum has led to a better understanding of the nature of the active sites in certain electron transfer proteins (such as the ferredoxins) and enzymes (such as nitrogenase)[16].

The characterisation of this wide range of cluster compounds represents a tribute to the power of modern spectroscopic[17] and X-ray crystallographic techniques[18]. These have provided a wealth of spectroscopic and structural data for the theoretical chemist to try and rationalise. Despite the great structural variety exhibited by cluster compounds, patterns have been recognised, which have prompted the development of generalisations which have been used in an attempt to unify the field.

From the earliest days of cluster chemistry, theoretical descriptions of the bonding in polyhedral clusters of the main group and transition metal elements have been divisible into three main types:

(i) Localised Bonding Treatments, where edge (2-centre-2-electron) or face (3-centre-2-electron) localised bonds are invoked to describe bonding in the cluster skeleton. These treatments have their origins in the ***Valence Bond*** approach developed by Heitler and London[19] and popularised by Pauling[20].

(ii) Delocalised Bonding Treatments, where the cluster valence electrons occupy ***Molecular Orbitals***[21] which are delocalised over the entire cluster framework. These molecular orbitals are often expressed as linear combinations of orbitals belonging to constituent fragments of the cluster, in order to make the bonding ideas more accessible to chemists.

(iii) Free Electron Treatments, where the cluster electrons are initially free to move over the surface of a sphere which represents the idealised cluster. In this limit the cluster wavefunctions are the solutions of the ***Schrödinger Equation*** for a particle on a sphere[22], with energies dependent on their nodal characteristics. Since the cluster atoms define a polyhedron which is not spherical, the orbitals and energies of the real cluster are estimated (from those at the spherical limit) by a perturbation theory analysis[22] which incorporates the actual symmetry of the cluster.

The popularities of these three approaches have waxed and waned as different researchers have espoused the merits of their particular favourites. The characterisation of increasingly complex cluster structures has tested the current bonding theories to the limit and has at times required their substantial revision. The mathematical methodologies of topology, graph theory and group theory have, at one time or another, been incorporated into each of the bonding models, with varying degrees of success.

B. Localised Cluster Bonding Models

I. Introduction to Localised Orbitals

The first clusters to be studied from a localised Valence Bond[19, 20] (VB) viewpoint were the boron hydride cage compounds $[B_nH_{n+m}]$ (m = 4 or 6)[2]. The VB approach describes bonding in terms of electron pair bonds involving the orbitals of two neighbouring atoms. It was apparent, however, that the polyhedral boron hydrides were "electron deficient" since they possessed more nearest neighbours (i.e. cluster edges) than bond pairs[23]. In 1949 Longuet-Higgins stated that any localised bonding description of the bonding in such molecules must include 3-centre-2-electron (3c–2e) bonds in addition to the classical 2-centre-2-electron (2c–2e) bonds[24].

Over the following five years, as more borane clusters were structurally characterised, localised bonding schemes involving 2- and 3-centre bonds (BH, BHB, BB and BBB) were applied to a variety of these molecules by Longuet-Higgins[25], Duffey[26] and Lipscomb[27]. The ***General Valence Orbital*** method of Torkington[28] was utilised by Duffey and Lipscomb to generate symmetry adapted hybrid orbitals at each cluster vertex, with the localised orbitals being formed by the interaction of these hybrids. The ***Equivalent Orbital*** methodology of Lennard-Jones[29] and Pople[30] was also used to transform a basis set of atomic orbitals into a set edge- or face-localised equivalent orbitals. In this way, the hypothetical tetrahedral species $[B_4H_4]$ was predicted to have 4 bond pairs, residing at the centres of its 4 triangular faces[26, 27].

II. The styx Model

Lipscomb has developed an elegant topological methodology, involving 2c- 2e and 3c-2e bonds, which has proved particularly useful for rationalising the structures of boranes of the type $[B_2H_{n+m}]$. His ***styx*** formalism[2a] starts from the premise that each boron atom participates in 4 bonds, of the following types:

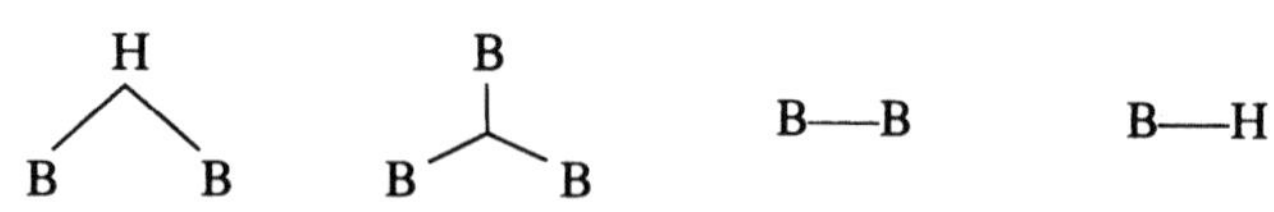

Description	3c–2e BHB	3c–2e BBB	2c–2e BB	2c–2e BH
Number	s	t	y	x

For a general borane $[B_nH_{n+m}]$, possessing n BH units and m *endo*-terminal or bridging hydrides, the total number (N_T) of bond pairs is:

$$N_T = s + t + y + x = 2n + (m/2) \quad (1)$$

(since each boron atom contributes 3 electrons and each hydrogen atom contributes 1 electron). By enumerating atoms, orbitals and electrons Lipscomb derived the following relationships[2a]:

$$x = m - s \qquad t = n - s \qquad 2y = s - x \quad (2)$$

and assigned each borane cluster a set of *styx* indices. This method successfully accounted for the geometries and charge distributions in the low symmetry borane cage compounds $[B_nH_{n+m}]$[2a, 31]. For the higher symmetry boranes, in particular the deltahedral (i.e. possessing exclusively triangular faces) anions $[B_nH_n]^{2-}$ shown in Fig. 1, the *styx* formalism led to an unworkable number of resonance forms because the bonding in such clusters can no longer be adequately described within a simple localised framework[23].

III. The Topological Equivalent Orbital Method

Kettle attempted to circumvent the resonance problems inherent in earlier localised orbital methods by allowing for interaction between the edge or face-localised cluster bonding orbitals. His ***Topological Equivalent Orbital*** methodology applied group theoreti-

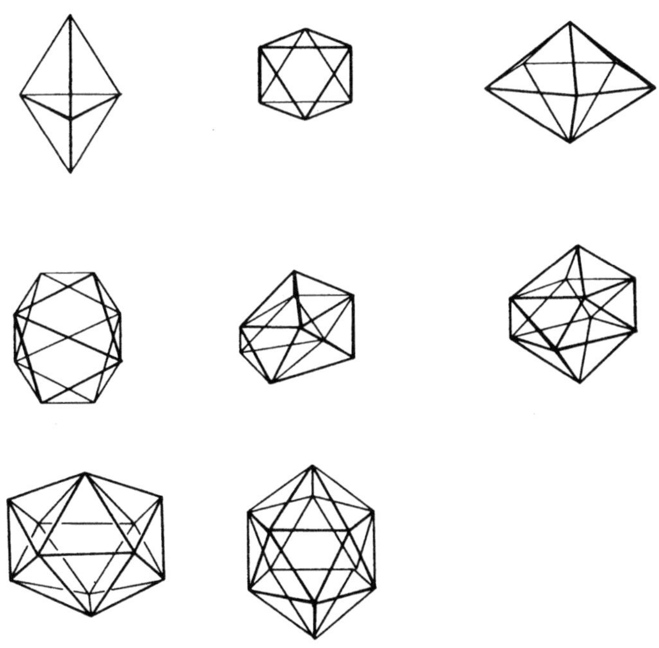

Fig. 1. Deltahedral geometries adopted by the borane anions $[B_nH_n]^{2-}$

cal labels to cluster orbitals which were constructed as linear combinations of localised orbitals, thereby constituting a compromise between purely localised bonding theories and (delocalised) molecular orbital theory. This analysis was successfully applied to the deltahedral borane anions[32] as well as the more open cage structures[33].

In the early 1960's the field of transition metal cluster chemistry expanded rapidly[34]. Some of the first metal clusters to be structurally characterised were the halide bridged clusters of the early transition metals[12], such as octahedral $[Mo_6Cl_8]^{4+}$ which possesses 8 face-bridging ligands[35]. Sheldon, in a Valence Bond treatment of this cluster, concluded that it possessed 12 discrete 2c–2e bonds, localised along the edges of the metal octahedron[36]. A subsequent qualitative hybrid orbital treatment by Duffey et al. supported this conclusion[37]. Cotton later criticised these Valence Bond treatments on the grounds that they implied the occurrence of significantly bent metal-metal bonds and unrealistically high formal oxidation states for the metal atoms[34a, 38]. Kettle subsequently demonstrated that the 12 cluster bonding orbitals of $[Mo_6Cl_8]^{4+}$ (from the Molecular Orbital treatment of Cotton and Haas[38]) are topological equivalent to symmetry adapted linear combinations of 12 edge-bonding localised orbitals[39]. He also showed that clusters of the type $[M_6Cl_{12}]^{2+}$ (M = Nb, Ta) (with 12 edge-bridging ligands[40]) possess 8 cluster bonding orbitals[38] which are topologically equivalent to linear combinations of 8 face-bonding (3c–2e) localised orbitals[39].

Kettle demonstrated that the bonding in small transition metal carbonyl clusters, such as triangular $[M_3(CO)_{12}]$ (M = Fe, Ru, Os) and tetrahedral $[M_4(CO)_{12}]$ (M = Co, Rh, Ir)[7], could also be discussed from a localised viewpoint. He partitioned the cluster valence electrons into those involved in terminal M–CO bonding (including π-back donation from the metal into the vacant π^* orbitals of CO), M–CO (bridging) bonding and those involved in the formation of 2-centre and 3-centre localised metal-metal bonding orbitals[41]. In a study of tetrahedral clusters of the main group and transition metal elements, he noted that the P_4 molecule[42] possesses 6 edge-localised (2c–2e) bonding orbitals[43]. Like $[Mo_6Cl_8]^{4+}$, P_4 is an example of an "electron precise" cluster because there is a one-to-one topological mapping between the edges of the cluster polyhedron and the cluster bond pairs.

IV. The Effective Atomic Number Rule

The total number of valence electrons in a cluster compound which may be adequately described, in Valence Bond terms, by edge-localised 2c–2e bonds can be calculated on the basis of the ***Inert Gas*** or ***Effective Atomic Number (EAN) Rule***[44] (such that each main group atom in the cluster skeleton has 8 valence electrons and each transition metal atom has 18 valence electrons associated with it). This connection between localised bonding descriptions and the EAN rule can be traced back to the work of Sidgwick[45] and Pauling[20] in the 1930's and '40's. The number of cluster valence electrons (N_e) for "electron precise" main group and transition metal clusters may be calculated from the following relationships:

main group clusters $N_e = 8M - 2E$

transition metal clusters $N_e = 18M - 2E$ (3)

where M is the number of atoms in the cluster skeleton and E is the number of edges of the cluster polyhedron.

The EAN rule predicts a valence electron count of 84 for an electron precise octahedral transition metal cluster. It was found, however, that although $[Mo_6Cl_8L_6]^{4+}$ is an 84 electron cluster (including six 2-electron-donating terminal ligands, L), all of the octahedral transition metal carbonyl clusters which had been characterised by the end of the 1960's possess 86 valence electrons (e.g. $[Rh_6(CO)_{16}]$[46]). Kettle, however, concluded that such clusters are in fact electron deficient, like the isostructural boranes, rather than electron rich as the EAN rule would seem to indicate[47]. Unfortunately, Kettle's Topological Equivalent Orbital treatment of the bonding in $[Rh_6(CO)_{16}]$ suffered from an inability to separate properly metal-metal and metal-carbonyl bonding orbitals so a direct comparison with the bonding in boranes was not possible. As will be shown in the next section, it was the delocalised bonding treatments by Wade and Mingos, based on the earlier analysis of Longuet-Higgins, which eventually resolved the problem of bonding in electron deficient borane and transition metal carbonyl clusters.

In the past 15 years, the synthesis of a series of 3-connected electron precise hydrocarbon clusters $[C_nH_n]$[5] (see Fig. 2 for examples) has renewed interest in localised bonding schemes, since the bonding orbitals in these clusters are topologically equivalent to edge-localised 2c–2e bond pairs in a Valence Bond sense. Schulman et al. performed *ab initio* calculations on cubane $[C_8H_8]$ utilising a basis set of 2c–2e localised C–C and C–H

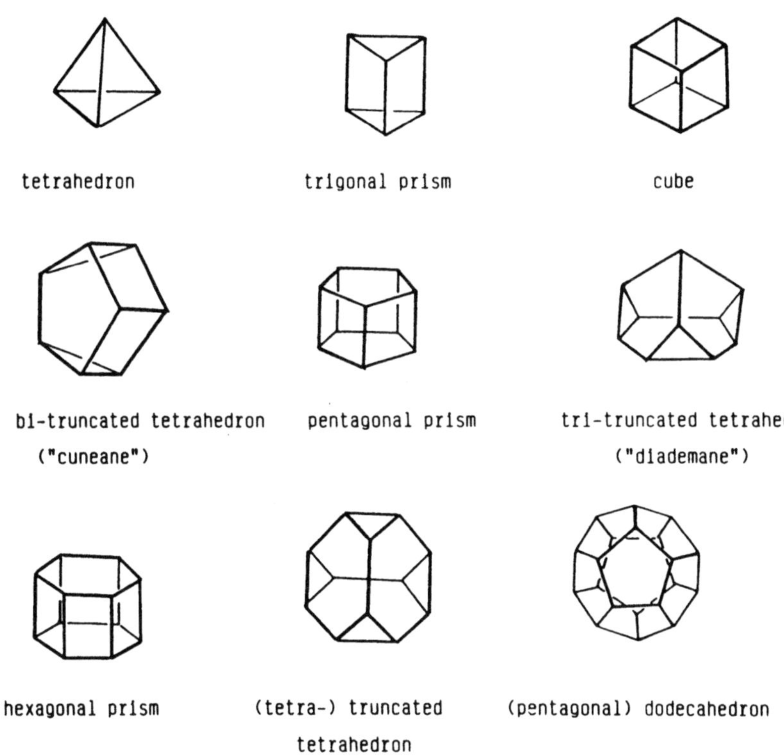

Fig. 2. 3-connected geometries adopted by the hydrocarbon clusters $[C_nH_n]$

bonding orbitals[48]. The resonance integrals (between localised C–C bonding orbitals) used in these calculations were quite large, thereby giving rise to a wide spectrum of cluster orbitals. Schulman's methodology, which was similar to that employed by Kettle, also included mixing between C–C and C–H bonding orbitals and led to an energy level ordering which was essentially the same as that obtained by delocalised LCAO-MO treatments[48]. A similar Equivalent Orbital method was applied by Heilbronner et al. to study the bonding in tetrahedrane $[C_4H_4]$[49].

V. Electron Rich Clusters: Bond Breaking and Skeletal Distortions

The formal addition of electron pairs to electron precise clusters generally results in the breaking of cluster bonds (one bond being broken for every electron pair added) with the resulting cluster being termed "electron rich". Mingos showed how this arises from a localised bonding viewpoint, for within such a framework each bond (edge) has associated with it one bonding and one antibonding orbital and the occupation of the latter leads to bond rupture[50]. He gave examples of electron precise (3-connected) and electron rich main group and transition metal clusters and showed how they are related by bond breaking until, ultimately, ring compounds are obtained. Gillespie subsequently developed this approach, showing that from a Valence Bond standpoint the addition of 2 electrons to a 2c–2e bond results in the cleavage of the bond and the formation of either 2 lone pairs or bonds to 2 additional ligands, as shown in Fig. 3[51]. This methodology was used to rationalise the structures of cluster, cage and ring compounds of the main group elements in terms of electrostatic repulsions between electron pairs (lone pairs and cluster-ligand bond pairs)[51] in an extension of the ***Valence Shell Electron Pair Repulsion Theory*** developed by Gillespie and Nyholm for main group AB_x molecules[52]. Figure 4 exemplifies these ideas by correlating the structures of some main group 8-atom cage and ring compounds with cuneane $[C_8H_8]$, from which structure they may be derived by cluster bond breaking. An exception to this rule of electron pair addition and bond rupture is found in the case of trigonal prismatic $[Te_6]^{4+}$ which possesses 9 edges and 10 electron pairs[53]. In this case, however, the additional electron pair leads to a stretching distortion parallel to the threefold rotational axis of the trigonal prism, rather than the

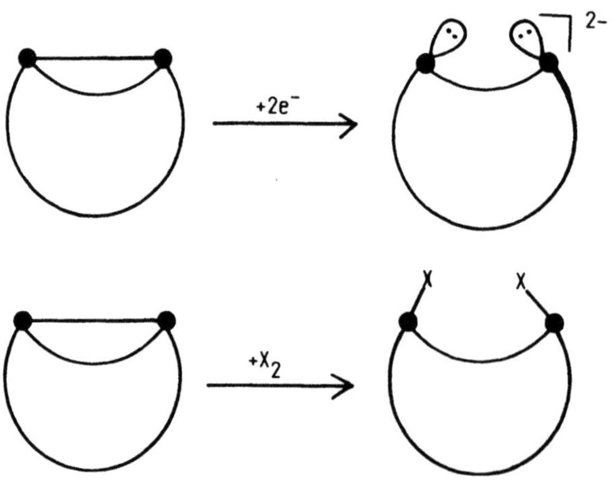

Fig. 3. Localised view of bond breaking processes

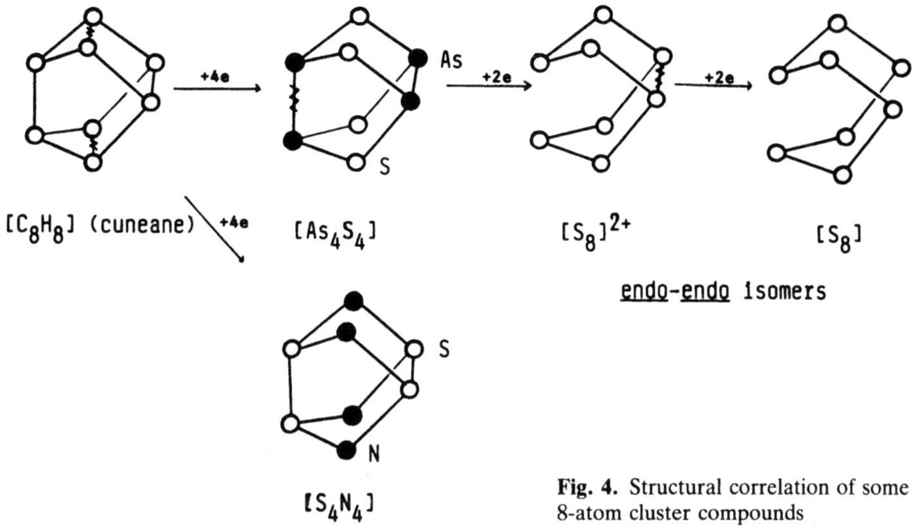

Fig. 4. Structural correlation of some 8-atom cluster compounds

localised breaking of one of the Te–Te bonds. Dahl has performed a series of elegant synthetic and structural studies on transition metal cluster compounds, which have investigated the geometrical consequences of adding and removing electrons[54]. His research has highlighted drawbacks in the localised bonding approach. For example, the formal addition of one electron, on going from $[Co_2Fe(CO)_9S]$ to $[Co_3(CO)_9S]$[54a, b], leads to an increase of all three metal-metal bond lengths. this indicates the population of an antibonding orbital which is delocalised over the entire triangular cluster, rather than localised between any two metal atoms[54].

VI. Recent Developments

McGlinchey and Tal have developed more topologically based arguments to rationalise and classify the structures adopted by electron rich clusters[55]. Their treatment follows on from the work of King who, over the past 20 years, has used topology, graph theory and group theory to analyse the bonding in electron deficient, electron precise and electron rich clusters of the main group and transition metal elements[56] as well as coordination compounds[57]. Teo has subsequently proposed a topological electron counting procedure for transition metal clusters[58]. His analysis is performed on the basis of the Effective Atomic Number rule and Euler's theorem (V + F = E + 2), by means of the formula:

$$\text{Number of cluster valence orbitals} = 8V + 2 - F + X \quad (4)$$

where V is the number of cluster vertices, F is the number of faces of the cluster polyhedron and X is the number of "additional" electron pairs required to match the true cluster electron count (thus X = 1 for an octahedral metal-carbonyl cluster). X reflects the many-centre character of the bonding in cluster compounds and is equal to 0 when the cluster is "electron precise". The method suffers, however, from the empirical way in which X is evaluated. Further recent applications of localised bonding models include studies by King[56c)] and Wade[59)] (applying Lipscomb's *styx* methodology) aimed at explaining the different arrangements of metal and boron atoms in the Δ-dodecahedral clusters $[Cp_4Co_4B_4H_4]$ (**1**)[60)] and $[Cp_4Ni_4B_4H_4]$ (**2**)[61)].

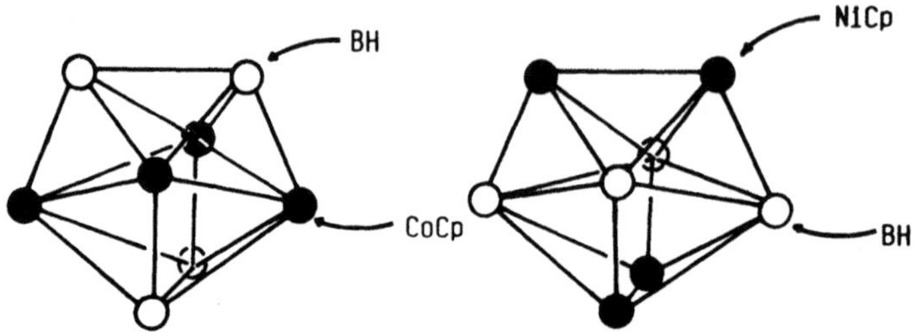

$[Cp_4Co_4B_4H_4]$

(**1**)

$[Cp_4Ni_4B_4H_4]$

(**2**)

C. Delocalised Cluster Bonding Models

I. Introduction

The problems associated with the application of localised bonding approaches, described in the previous section, can be conveniently circumvented within the ***Molecular Orbital (MO)*** framework[21)], where the cluster molecular orbitals ($\Psi_{cluster}$) may be expressed as ***linear combinations of atomic orbitals (LCAO's)*** (ϕ_i)[62)]:

$$\Psi_{cluster} = N \sum_i c_i \phi_i \tag{5}$$

where N is a normalisation coefficient and c_i is the coefficient of the ith atomic orbital.

II. Borane Clusters

1. Early Studies

The pioneering work of Longuet-Higgins, on the highly symmetrical deltahedral boranes[63)], led to the prediction of icosahedral $[B_{12}H_{12}]^{2-}$ some years before its actual synthesis[64)]. He showed that each BH fragment contributes 3 orbitals (one radial sp hybrid and 2 tangential p orbitals) towards the 13 skeletal bonding molecular orbitals of this electron deficient cluster[64)]. Lipscomb and Hoffmann performed a series of semi-empirical MO calculations on the polyhedral boranes and carboranes[65)]. They used calculated Mulliken overlap populations[66)] and charge distributions to explain bond length and reactivity trends in these "superaromatic" cluster species. Throughout the 1960's Muetterties, Guggenberger and their coworkers used Extended-Hückel calculations[65a, 67)] to evaluate the total energies of borane clusters in order to rationalise their structures in solution and in the solid state[68]. More recently, less empirical (and even *ab initio*) molecular orbital methods have proved very useful in rationalising the wealth of data arising from experimental investigations of physical properties of specific boranes and carboranes[69].

2. Structure-Electron Count Correlations

At the beginning of the 1970's, although many molecular orbital calculations had been performed on individual borane and carborane clusters, no generalisations had been made which would result in a theoretical unification of the field. Initially most boranes were thought to adopt structures based on fragments of an icosahedron, but it later became apparent, as more borane clusters were structurally characterised, that other deltahedrally based structures were quite common[70]. Williams first elucidated a struc-

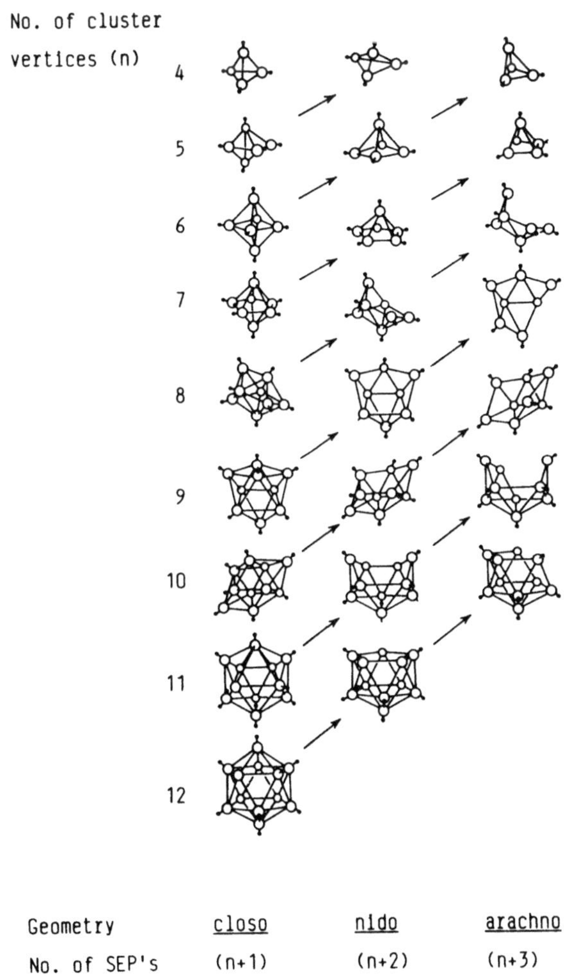

Fig. 5. Structure-skeletal electron pair correlation for *closo, nido* and *arachno* boranes

tural pattern which was based on the recognition of three main structural types[71]. The deltahedral boranes $[B_nH_n]^{2-}$ and carboranes (e.g. $[C_2B_{(n-2)}H_n]$) were given the prefix *closo* (from the Greek for "closed"). The *nido* (from the Greek for "nest-like") structures were defined as being derived from *closo* clusters by the removal of one (BH) group (i.e. one cluster vertex) and boranes of the type $[B_nH_{n+4}]$ were found to belong to this class of cluster. The *arachno* (from the Greek for "web-like") structures were defined as being derived from *closo* clusters by the removal of two (BH) cluster vertex units and the boranes $[B_nH_{n+6}]$ were found to belong to this class.

In a seminal account, Wade exposed the relationship between the structural generalisations of Williams and the number of delocalised (cluster-bonding) ***skeletal electron pairs*** (SEP's) possessed by the cluster polyhedra[72a]. He recognised that the (n + 1) skeletal electron pairs characteristic of the *closo* boranes $[B_nH_n]^{2-}$ were retained in the *nido* and *arachno* derivatives, $[B_{n-1}H_{(n-1)+4}]$ and $[B_{n-2}H_{(n-2)+6}]$[72]. In *nido* and *arachno* boranes $[B_xH_{x+y}]$ there are x(BH) units and y additional hydrogen atoms which occupy *endo*-terminal or bridging positions around the open face of the cluster. These hydrogen atoms serve to lower the formal negative charge carried by the cluster and to stabilise those cluster molecular orbitals which have large amplitudes around the open face.

Wade pointed out that a convenient way of looking at the *closo, nido, arachno* pattern is to consider the addition of successive pairs of electrons to an n-atom *closo* cluster with (n + 1) SEP's[72a]. This results in an opening out of the structure to yield *nido* and *arachno* structures, characterised by (n + 2) and (n + 3) SEP's respectively. Rudolph subsequently produced a correlation diagram (reproduced in Fig. 5) which combined the approaches of decapping (i.e. the removal of cluster vertices, also known as the ***Debor Principle***) and electron pair addition[73].

III. Transition Metal Clusters

1. Early Studies

As alluded to earlier, a seminal theoretical study in the field of transition metal cluster chemistry was conducted by Cotton and Haas on halide bridged clusters of the type $[M_6(\mu_3-X)_8X_6]^{2-}$ (face-bridged octahedron; M = Mo, W; X = Cl, Br) and $[M_6(\mu_2-X)_{12}X_6]^{4-}$ (edge-bridged octahedron; M = Nb, Ta; X = Cl, Br)[38]. They discovered that the face-bridged cluster (with 84 cluster valence electrons) possesses 12 SEP's while the edge-bridged cluster (with 76 cluster valence electrons) possesses 8 SEP's. These skeletal electron pairs were found to reside in molecular orbitals composed of metal d orbitals and the relative numbers of SEP's were rationalised in terms of the relative disposition of the bridging ligands and the lobes of the metal atom d orbitals[38]. We have recently extended this work to discuss the bonding in *nido* and *arachno* derivatives of both the face- and the edge-bridged octahedral clusters[74].

Although the above mentioned face- and edge-halide-bridged clusters are characterised by 84 and 76 valence electrons respectively, it was apparent by the early 1970's that the great majority of octahedral transition metal clusters possess 86 electrons. Braterman rationalised the 86 cluster valence electron count for octahedral metal carbonyl clusters, such as $[Co_6(CO)_{16}]$[75], $[Ru_6C(CO)_{17}]$[76] and $[Co_6(CO)_{14}]^{4-}$[77] in terms of localised M–CO bonds and delocalised skeletal molecular orbitals with predominantly metal d orbital character[78]. Drawing heavily on the above mentioned molecular orbital calculations of Cotton and Haas, he suggested that $[Co_6(CO)_{16}]$, for example, possessed 7 bonding skeletal molecular orbitals derived from the metal d_{z^2}, d_{xz} and d_{yz} orbitals, with all of the linear combinations of the d_{xy} and $d_{x^2-x^2}$ orbitals taken to be roughly non-bonding and, therefore, occupied.

2. *Polyhedral Skeletal Electron Pair Theory*

Wade clarified matters considerably by recognising that the bonding pattern observed for *closo, nido* and *arachno* boranes is reproduced by metal carbonyl clusters[72]. He noted that $[H_2Ru_6(CO)_{18}]$[79] and $[Fe_5C(CO)_{15}]$[80] constituted a related pair of octahedral *(closo)* and square pyramidal *(nido)* clusters, both possessing 7 skeletal bonding molecular orbitals and that they were, therefore, analogous to $[B_6H_6]^{2-}$ and $[B_5H_9]$[72a]. This conclusion depended on the assumption that each metal tricarbonyl fragment in these clusters (like the BH and CH^+ fragments in boranes and carboranes) contributes 3 orbitals and 2 electrons for skeletal bonding. In several subsequent papers Wade[81] and Mingos[50] extended these ideas to electron precise and electron rich clusters and recognised the important bonding similarities between certain main group and transition metal fragments. The term ***Polyhedral Skeletal Electron Pair Theory*** was introduced by Mason, Thomas and Mingos to cover these theoretical correlations of cluster structure with the number of electrons involved in skeletal bonding[82]. Subsequent Extended Hückel calculations by Evans confirmed the applicability of PSEPT to a wide range of cluster geometries[83].

Extended Hückel calculations by Mingos, on the octahedral cluster $[Co_6(CO)_{14}]^{4-}$, showed the occurrence of wide bands of orbitals formed from the valence 4s and 4p atomic orbitals on the metal atoms while the 3d-band was calculated to be quite narrow on account of the small 3d–3d overlap integrals[84]. The metal valence s and p orbitals were found to give rise to 11 strongly antibonding cluster molecular orbitals which are "inaccessible" due to their high energies. They are also unsuitable for metal-ligand bonding due to being inwardly hybridised. Mingos realised that these 11 orbitals have identical nodal and symmetry properties to those of the high-lying antibonding orbitals of $[B_6H_6]^{2-}$. The valence electron counts of these two octahedral clusters may be related as shown below:

	$[B_6H_6]^{2-}$	$[Co_6(CO)_{14}]^{4-}$	
Valence electron count:	$2 \times \{(4 \times 6) - 11\} = 26$	$2 \times \{(9 \times 6) - 11\} = 86$	(6)

(since main group atoms possess 4 valence orbitals and transition metals possess 9 valence orbitals).

Lauher, subsequently, performed Extended Hückel calculations on a wide range of n-atom *closo* transition metal clusters and confirmed the presence of $(2n - 1)$ high-lying "inaccessible" antibonding orbitals, leaving a total of $\{9n - (2n - 1)\} = (7n + 1)$ occupied cluster valence MO's and an electron count of $(14n + 2)$[85]. Lauher's calculations were made on bare metal clusters and the introduction of ligands was considered to constitute only a small perturbation. Calculations on *nido* and *arachno* transition metal clusters showed them to be characterised by $(14n + 4)$ and $(14n + 6)$ valence electrons respectively[85)]. The difference in electron count between isostructural (n-atom) main group and transition metal clusters is 10 n, corresponding to the occupation of 5 d orbitals per metal atom. These orbitals are either non-bonding (i.e. localised exclusively on the metals) or mix with the s and p orbitals for form inwardly hybridised orbitals which are involved in metal-metal bonding and outwardly hybridised orbitals which are involved in metal-ligand bonding[50, 72)].

3. *Isolobal Relationships*

As mentioned above, Wade and Mingos recognised that the underlying reason for structural similarities between main group and transition metal carbonyl clusters was that the constituent fragments contributed the same number of electrons and orbitals for cluster bonding and that these orbitals had similar nodal properties[50, 72)], Halpern had previously noted the electronic equivalence of certain metal fragments and organic groups[86)]. Similarities in the reactivity of alkyl radicals and $[Co(CN)_5]^{3-}$, as well as between carbenes and $d^8 - ML_4$ complexes, illustrated this equivalence. Later Dahl draw attention to similarities in the bonding modes of arsenic atoms and conical $Co(CO)_3$ fragments in tetrahedral clusters such as $[As_3Co(CO)_3]$ and $[As_2Co_2(CO)_6]$[87)]. Similar features were noted by Ellis[88)]. Rudolph rationalised the structures of the hetero-boranes by considering isoelectronic main group fragments[73)]. The fact that CH and BH^- are isoelectronic explains why $[C_2B_{(n-2)}H_n]$ and $[B_nH_n]^{2-}$ are isostructural. Similarly, the icosahedral cluster $[B_{11}H_{11}S]$[89)] is isoelectronic with the isostructural $[B_{12}H_{12}]^{2-}$ molecule, if the sulphur atom is assumed to possess one lone pair instead of a B–H bond pair and 4 electrons in 3 orbitals which are used for cluster bonding (hence making it isoelectronic with BH^{2-}). This approach may be extended to the bare metal (and non-metal) cluster and cage compounds of the post transition elements[4)]. As early as 1968, Corbett had pointed out that the trigonal bipyramidal cluster $[Bi_5]^{3+}$ possesses 6 skeletal bond pairs, assuming one lone pair on each bismuth atom[90)] (it is therefore isoelectronic with the isostructural $[C_2B_3H_5]$).

Semi-empirical molecular orbital analyses, by Hoffmann and coworkers, of the frontier molecular orbitals of metal carbonyl fragments confirmed that there were definite similarities between the numbers, energies, shapes and nodal characteristics of these orbitals and those of main group fragments EH_n[91)]. The term "*isolobal*" was introduced to describe this relationship[89b)] and the "*isolobal* analogy" has subsequently been elegantly formulated (for a variety of fragments) by Hoffmann[92)]. Figure 6 illustrates the frontier MO's of some *isolobal* hydrocarbon, CH_n, and ML_m fragments. The *isolobal* analogy has proved very useful in accounting for structural similarities between main group and transition metal cluster compounds. For example the *isolobal* nature of CR(R = alkyl or aryl) and conical $Co(CO)_3$ fragments, both presenting 3 orbitals and 3 electrons for cluster bonding, explains the occurrence of a series of tetrahedral clusters $[(Co(CO)_3)_n(CR)_{4-n}]$ (n = 1 – 4) which all formally possess 6 SEP's[92)]. Further examples of *isolobal* main group and transition metal fragments are given in Table 1 (which is adapted from one drawn up by Stone[93)].

Figure 6 shows that an $M(CO)_3$ fragment has 6 electrons in M–CO (σ) bonding orbitals and 6 electrons in a roughly non-bonding set of metal d orbitals. These orbitals are analogous to the t_{2g} orbitals of the octahedral complex from which the fragment is formally derived[92)] (as Wade and Mingos had previously concluded[50, 72)]). An alternative

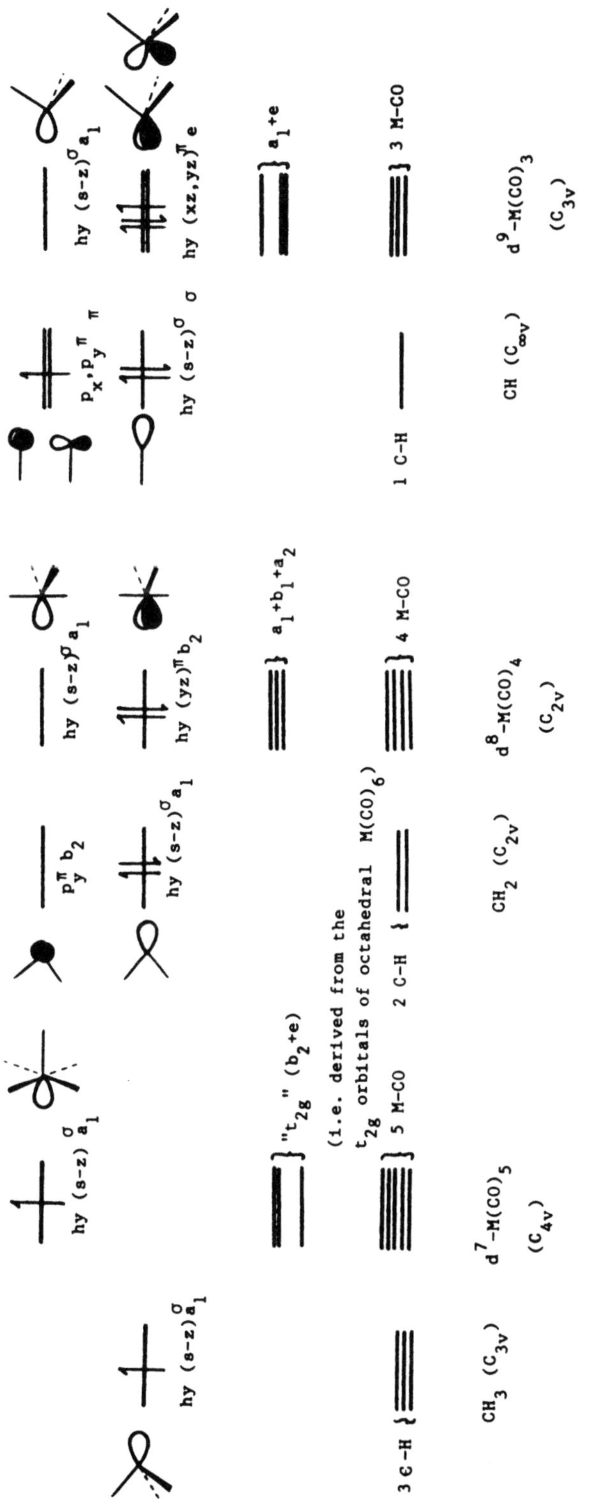

Fig. 6. Examples of *isolobal* CH_n and ML_m fragments

expression for the number of valence electrons (N_e) of an n-atom transition metal cluster may, therefore, be written:

$$N_e = 12n + 2S \quad (7)$$

where S is the number of skeletal electron pairs. This equation can then be used to calculate the number of SEP's of a cluster of known formula (and hence known valence electron count).

The semi-empirical arguments upon which the *isolobal* analogy is based have recently been corroborated by Fehler and DeKock[94], who performed LCAO-MO-SCF Fenske-Hall calculations on the *nido* square pyramidal clusters $[Fe(CO)_3B_4H_8]$[95], $[CoCpB_4H_8]$[96] and $[B_5H_9]$. In contrast to an earlier Self Consistent Charge-Extended Hückel study by Brint and Spalding[97], their calculations showed that the 3, supposedly *isolobal,* fragments ($Fe(CO)_3$, CoCp and BH) do indeed contribute 3 orbitals (one radial nodeless orbital and two tangential singly noded orbitals) and 3 electrons for cluster bonding. It should be noted, however, that Woolley has stated that, although the *isolobal* analogy does successfully rationalise transition metal cluster geometries and is a useful tool for the synthetic chemist[93], the detailed electronic structures of main group and transition

Table 1. Examples of *isolobal* main group and transition metal fragments[a]

CH_3	CH_2	CH
H	S (2 lone pairs)[b]	P, As (1 lone pair)[c]
$M(CO)_5$ (Mn, Tc, Re)	$M(CO)_4$ (Fe, Ru, Os)	$M(CO)_3$ (Co, Rh, Ir)
$Fe(CO)_2Cp$	Rh(CO)Cp	NiCp
$Mo(CO)_3Cp$	$Re(CO)_2Cp$	$W(CO)_2Cp$
	Cr(CO)(NO)Cp	$[Mn(CO)_2Cp]^-$
$Co(CO)_4$	$Cr(CO)_5$	$Re(CO)_4$
$PtH(PR_3)_2$[d]	$Pt(PR_3)_2$[e]	
ZnCp	CuCp	RhBz
$Au(PR_3)$	$IrCl(CO)_2$	$Re(CO)_3Br_2$
$Rh(PR_3)_2(\eta^5 - C_2B_9H_{11})$	$Fe(CO)_2(\eta^5 - C_2B_9H_{11})$	$Mn(CO)_2(\eta^5 - C_5B_9H_{11})$
	$TaMeCp_2$	$TaCl(PR_3)_2Cp$
CH_3^+	CH_2^+	CH^+
BH_3	BH_2, P (2 lone pairs)	BH, Sn (1 lone pair)
$M(CO)_5$ (Cr, Mo, W)	$M(CO)_4$ (Mn, Tc, Re)	$M(CO)_3$ (Fe, Ru, Os)
$Mn(CO)_2Cp$	Fe(CO)Cp	MCp (Co, Rh, Ir)

[a] *Isolobal* relationships are not one-to-one mappings, so the same metal fragment may (depending on what it is coordinated to) be effectively *isolobal* to more than one main group fragment. The frontier orbital properties of metal fragments (ML_n) depend critically on the nature of the ligands (L) and the geometry of the fragment. Although metal fragments are generally regarded as being derived from octahedral complexes, other "starting" geometries are possible (e.g. a d^1-ML_8 fragment, with a geometry based on the $[ReH_9]^{2-}$ molecule, is isolobal with CH_3; Hoffmann, R.: Angew. Chem. Int. Edn. Engl. *21,* 711 (1982).

[b] i.e. a 2-coordinated sulphur atom.

[c] i.e. an n-connected phosphorous, arsenic atom ($n > 2$).

[d] T-shaped ML_3 fragment.

[e] Bent ML_2 fragment.

metal clusters are quite different[98]. In particular, he has criticised some of the early semi-empirical treatments (notably that of Mingos[84]) for underestimating the d-band width in transition metal clusters and underplaying the important contribution to metal-metal bonding made by the d orbitals. Woolley has concluded that d orbital interactions dominate the metal-metal bonding in transition metal cluster compounds[98, 99].

4. *The Capping Principle*

In the 1970's, the only known large transition metal clusters ($n > 6$) with (borane-like) deltahedrally based (*closo, nido* and *arachno*) skeletal geometries were those possessing interstitial main group atoms (see Table 2 for examples)[100]. The majority of large metal clusters were found, instead, to have structures corresponding to capped polyhedra (generally capped tetrahedra and octahedra). Mingos introduced the ***Capping Principle***[50, 82, 101] to deal with these clusters (such as the capped octahedron $[Rh_7(CO)_{10}]^{3-}$ [102] and the bicapped tetrahedron $[Os_6(CO)_{16}]$ [82]). He demonstrated that capping a face of a polyhedron leads to no change in the number of skeletal bonding MO's. This is because the frontier orbitals of the capping fragment are matched in symmetry by orbitals of the parent cluster which are already bonding[101]. Capping with a conical $M(CO)_3$ fragment, therefore, leads to an increase in cluster valence electron count of 12 (since the fragment has 12 electrons in non-bonding and metal-ligand bonding orbitals). Thus, capped deltahedral clusters are characterised by $\{(14n + 2) + 12m\}$ valence electrons, where n is the number of metal atoms in the parent deltahedron and m is the number of capping

Table 2. *Closo, nido* and *arachno* transition metal clusters (based on deltahedra with more than 6 skeletal atoms) possessing interstitial main group atoms

Compound	Ref.	Metal Skeletal Geometry	No. of SEP's
$[Os_5C(CO)_{16}]$	(a)	*arachno*-pentagonal bipyramid	8 (n + 3)
$[Co_8C(CO)_{18}]^{2-}$	(b)		9 (n + 1)(*)
$[Ni_8C(CO)_{16}]^{2-}$	(c)	square-antiprism *(arachno)*	11 (n + 3)
$[Ni_9C(CO)_{17}]^{2-}$	(c)	capped square antiprism *(nido)*	11 (n + 2)
$[Rh_9P(CO)_{21}]^{2-}$	(d)		
$[Rh_{10}S(CO)_{22}]^{2-}$	(e)	*closo*-bicapped square antiprism	11 (n + 1)
$[Rh_{10}E(CO)_{22}]^{3-}$	(f)		
$[Rh_{12}Sb(CO)_{27}]^{3-}$	(g)	*closo*-icosahedron	13 (n + 1)

(a) Johnson, B. F. G., Lewis, J., Nicholls, J. N., Nelson, W. J. H., Puga, J., Raithby, P. R., Vargas, M. D.: J. Chem. Soc., Dalton Trans. 2447 (1983).
(b) Albano, V. G., Chini, P., Ciani, G., Martinengo, S., Sansoni, M.: J. Chem. Soc., Dalton Trans. 463 (1978).
(c) Longoni, G., Ceriotti, A., Della Pergola, R., Manassero, M., Perego, M., Piro, G., Sansoni, M.: Phil. Trans. R. Soc. Lond. *A308,* 47 (1982).
(d) Vidal, J. L., Walker, W. E., Pruett, R. L., Schoening, R. C.: Inorg. Chem. *18,* 129 (1979).
(e) Ciani, G., Garlaschelli, L., Sironi, A., Martinengo, S.: J. Chem. Soc., Chem. Commun. 563 (1981).
(f) (E = P or As) Vidal, J. L., Walker, W. E., Schoening, R. C.: Inorg. Chem. *20,* 238, 243 (1981).
(g) Vidal, J. L., Troup, J. M.: J. Organomet. Chem. *213,* 351 (1981).
* This cluster has a *closo* electron count, but the skeletal geometry lies in between the square antiprism *(arachno)* and the Δ-dodecahedron *(closo).*

metal fragments. This leads to electron counts of $14 n_t$ and $(14 n_t - 2)$ for monocapped and bicapped deltahedra respectively (where n_t is the total number (n + m) of metal atoms in the cluster).

Evans and Mingos have subsequently extended the Capping Principle to include edge-bridging C_{2v} ML_4 fragments[103] Hoffmann had earlier shown that d^8-ML_4 (C_{2v}) fragments are *isolobal* with carbene (CH_2) units[92], which explains their tendency to form ring compounds (e.g. $[Os_3(CO)_{12}]$[7]) and planar raft clusters (e.g. $[Os_6(CO)_{17}(P(OMe)_3)_4]$ (**3**))[104]. As Fig. 6 shows, an $Os(CO)_4$ fragment has 14 electrons in non-bonding (metal) and metal-ligand bonding orbitals, so that bridging the edge of a cluster with m of these groups leads to an increase in the cluster valence electron count of 14 m. For example, edge-bridged deltahedra are characterised by {(14 n + 2) + 14 m} electrons[103]. This again depends on the fact that the frontier orbitals of the bridging group generally interact with the bonding orbitals of the parent cluster. Mingos has shown, however, that when there are a large number of capping or bridging groups, some linear combinations may be generated which have the correct energy and symmetry to interact strongly with antibonding orbitals of the parent polyhedron. This generates additional cluster bonding MO's, thereby leading to an increase in the number of SEP's[101, 103]. The electronic requirements of main group and transition metal cluster and ring compounds (obtained from the Polyhedral Skeletal Electron Pair Theory) are summarised in Table 3 (adapted from Ref. 103).

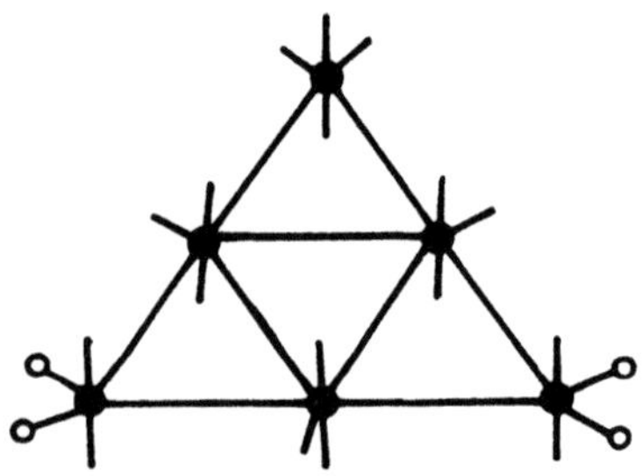

$[Os_6(CO)_{17}\{P(OMe)_3\}_4]$

(3)

Table 3. Summary of cluster valence electron counts from the Polyhedral Skeletal Electron Pair approach

Polyhedra	Main Group	Example	Transition Metal	Example
Deltahedra *(closo)*	4 n + 2	$[B_nH_n]^{2-}$	14 n + 2	$[Os_5(CO)_{16}]$
nido	4 n + 4	$[B_nH_{n+4}]$	14 n + 4	$[Ru_5C(CO)_{15}]$
arachno	4 n + 6	$[B_nH_{n+6}]$	14 n + 6	$[Os_4H_3I(CO)_{12}]$
3-connected polyhedra	5 n	$[C_nH_n]$	15 n	$[Ir_4(CO)_{12}]$
Rings	6 n	$[C_nH_{2n}]$	16 n	$[Os_3(CO)_{12}]$
Capped deltahedra (m capping groups)	4 n + 2 + 2 m	–	14 n + 2 + 12 m	$[Os_6(CO)_{18}]$
Edge-bridged deltahedra (m edge-bridging groups)	4 n + 2 + 4 m	–	14 n + 2 + 14 m	–

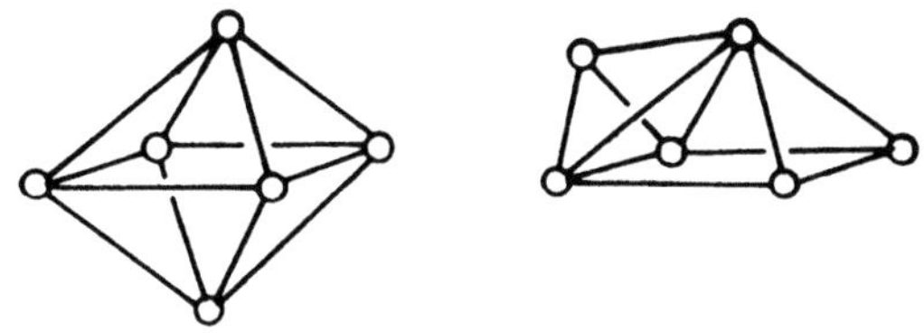

$[Os_6(CO)_{18}]^{2-}$ $[H_2Os_6(CO)_{18}]$

Fig. 7. Alternative structures for 86 electron Os_6 clusters

Combination of the Capping Principle and the Debor Principle leads to some interesting structural possibilities. Figure 7 depicts two isoelectronic Os_6 clusters which have different cluster geometries. The *closo* octahedral cluster $[Os_6(CO)_{18}]^{2-}$ has 86 cluster valence electrons ($86 = 14\,n + 2$, $n = 6$) as does the *Δ*-capped-*nido* square pyramidal cluster $[H_2Os_6(CO)_{18}]$ ($86 = (14\,n + 4) + 12\,m$, $n = 5$, $m = 1$)[105]. Diagrams correlating skeletal geometries and numbers of cluster valence electrons, making use of the Capping and Debor principles, have been constructed by Evans and Mingos[103] and McPartlin[106]. An example of such a correlation diagram is reproduced in Fig. 8.

5. *Platinum Clusters*

For the later transition metals, the d-p promotion energies are large, which in mononuclear compounds leads to a large number of stable 16 and 14 electron complexes. Over the past ten years a number of platinum clusters, usually with phosphine ligands, have been structurally characterised[10]. These clusters (examples of which are listed in Table 4) generally possess fewer electrons than would be expected on the basis of the Polyhedral Skeletal Electron Pair Theory. Evans and Mingos have presented a detailed analysis of the bonding in clusters of this type, concluding that the high energy of the atomic p orbitals and the deficiency of ligands, leads to some of the cluster MO's with predominantly p^{π} orbital character being unoccupied. This is in addition to those orbitals previously described as "inaccessible" due to being strongly metal-metal antibonding[107]. Their

Table 4. Examples of platinum cluster compounds[a]

Skeletal Geometry	Example	Electron Count	
Triangle	$[Pt_3(CO)_3(PR_3)_3]$	42	$(48)^b$
	$[Pt_3(CO)_3(PR_3)_4]$	44	
Tetrahedron	$[Pt_4H_8(PR_3)_4]$	56	(60)
Butterfly	$[Pt_4(CO)_5(PR_3)_4]$	58	(62)
Trigonal Bipyramid	$[Pt_5H_8(PR_3)_5]$	68	(72)
Edge-bridged Tetrahedron	$[Pt_5(CO)_6(PR_3)_4]$	70	(74)
Trigonal Prism	$[Pt_6(CO)_{12}]^{2-}$	86	(90)

[a] Mingos, D. M. P., Wardle, R. W. M.: Transition Met. Chem. *10,* 441 (1985).

[b] Numbers in parentheses are valence electron counts for clusters of the manganese, iron and cobalt groups.

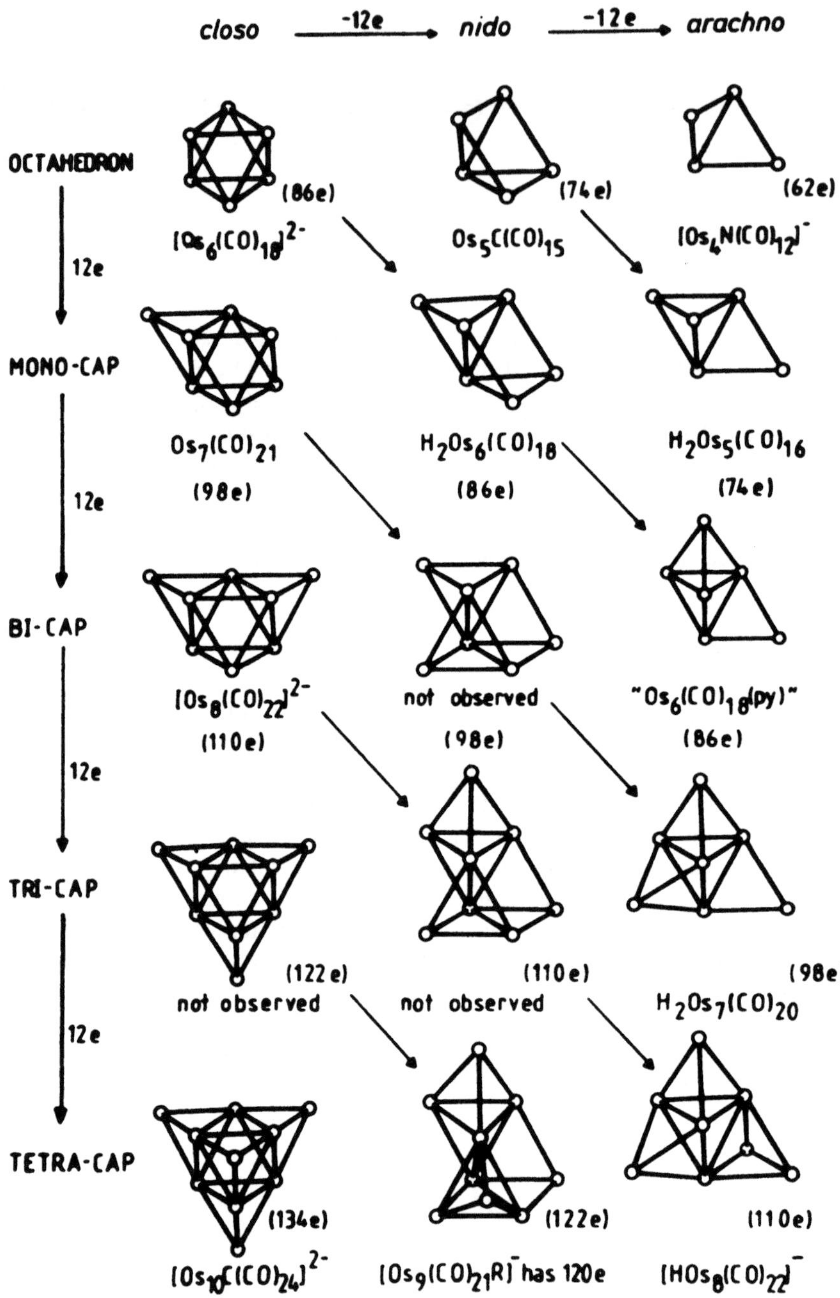

Fig. 8. Correlation of structure with valence electron count for transition metal cluster, utilising the Capping and Debor principles

analysis involved the extension of the *isolobal* analogy to include PtL_2 fragments, the bonding capabilities of which Mingos had studied several years earlier in his theoretical treatment of platinacarboranes[108].

6. Gold Clusters

Relative to those of the earlier transition metals, the d orbitals of the coinage metals (Cu, Ag and Au), are contracted and low lying. Thus, they make only a small contribution to cluster bonding[109]. Since the s-p promotion energy is very large[110], the valence p orbitals also do not play a large part in metal-metal bonding. Instead, bonding occurs primarily through the valence s orbitals[109]. A large number of gold-phosphine clusters have been structurally characterised[11] and several copper clusters are known (usually with bridging hydride, alkyl or aryl ligands supporting the weak metal-metal bonding)[111], while cluster compounds of silver are rare[112].

Evans and Mingos have shown that the $Au(PR_3)$ fragment has a low-lying singly occupied sp hybrid orbital (with predominantly s character) and two empty tangential p orbitals, which are too high-lying to become involved in cluster bonding[113]. The fragment is, therefore, effectively *isolobal* with, for example, H, CH_3 and $Mn(CO)_5$, an analogy which has been synthetically exploited by Lewis, Stone and others[114].

There are a number of gold clusters with the general formula $[Au_m(PR_3)_m]^{x+}$ (see Fig. 9 for examples) whose electron counts are critically dependent on the cluster geometry[11]. Evans and Mingos have concluded that the occupied skeletal MO's, in these clusters, are low lying (nodeless or singly noded) linear combinations of Au sp hybrids orbitals[113]. It has been noted that many of the higher nuclearity gold clusters possess structures consisting of a gold atom in the centre of a cage, which is a fragment of an icosahedron[109]. This led Mingos to predict the possibility of synthesising a centred-icosahedral $[Au_{13}]$ cluster[109, 115], a prediction which was subsequently borne out with the characerisation of $[Au_{13}(PMe_2Ph)_{10}Cl_2]^{3+}$ [116].

High nuclearity gold clusters have the general formula $[Au\{Au(PR_3)\}_m]^{x+}$ and may be divided into two sub-classes, according to whether the outer gold atoms adopt a

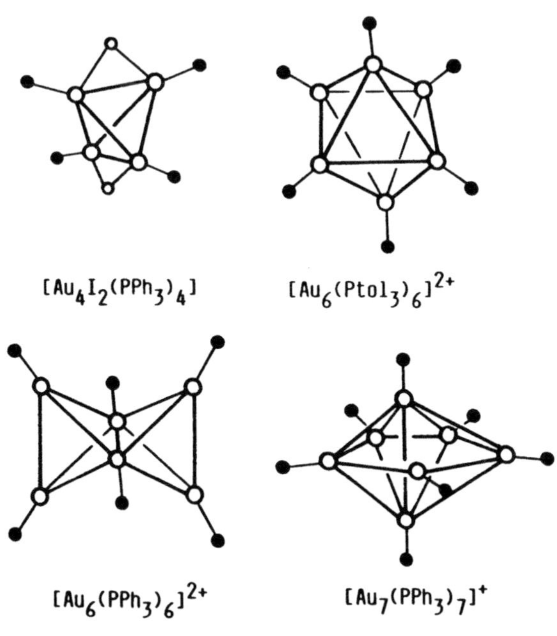

Fig. 9. Examples of gold clusters of the form $[Au_m(PR_3)_m]^{x+}$

spherical or toroidal topology[117]. "Spherical" clusters (see Fig. 10a for examples) are characterised by (12 m + 18) cluster valence electrons. Thus, all 9 valence orbitals of the central gold atom are involved in bonding with the radial sp hybrids of the outer AuL groups, which each contribute 2 metal-ligand bonding and 10 non-bonding (d) electrons to the total count. "Toroidal" clusters (see Fig. 10b for examples) are characterised by (12 m + 16) cluster valence electrons, since one of the p orbitals on the central gold atom finds no low-lying symmetry-matched cluster orbital[117].

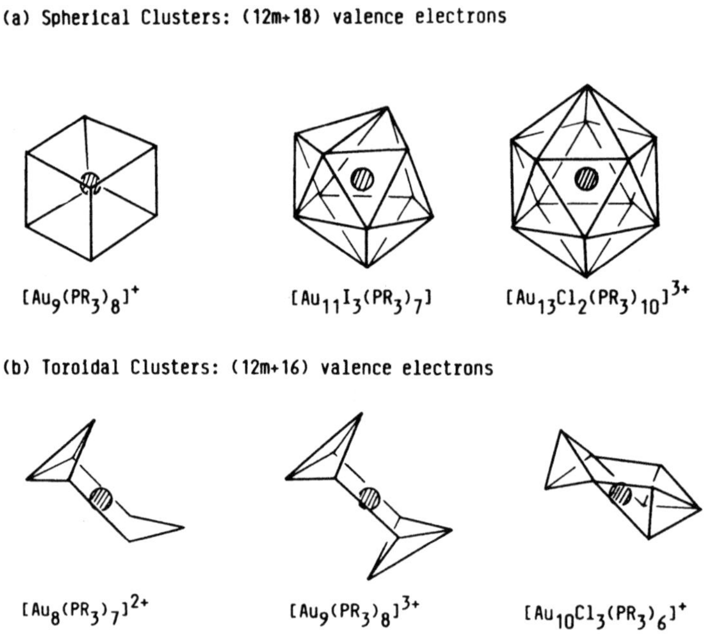

Fig. 10. Examples of high nuclearity gold clusters, $[Au\{Au(PR_3)\}_m]^{x+}$, with spherical and toroidal topologies

IV. High Nuclearity Transition Metal Clusters

1. The Principle of Polyhedral Fusion

In recent years a large number of high nuclearity transition metal-carbonyl clusters have been synthesised, most of which have structures corresponding to the condensation of smaller tetrahedral, octahedral and trigonal prismatic cluster fragments[7b]. Mingos has recently developed a methodology which has enabled him to formulate rules to account for the cluster valence electron counts in these molecules. In this ***"Polyhedral Fusion"*** formalism, the condensed cluster is regarded as the product of two polyhedra sharing a vertex, edge or face[118, 119]. In this way Mingos derived the rule that ***"the total electron count in a condensed cluster is equal to the sum of the electron counts for the parent polyhedra (A) and (B) minus the electron count (Δ_c) characteristic of the shared unit (atom, pair of atoms etc.)"***[118]. Thus, if polyhedron (A) is characterised by a electrons and

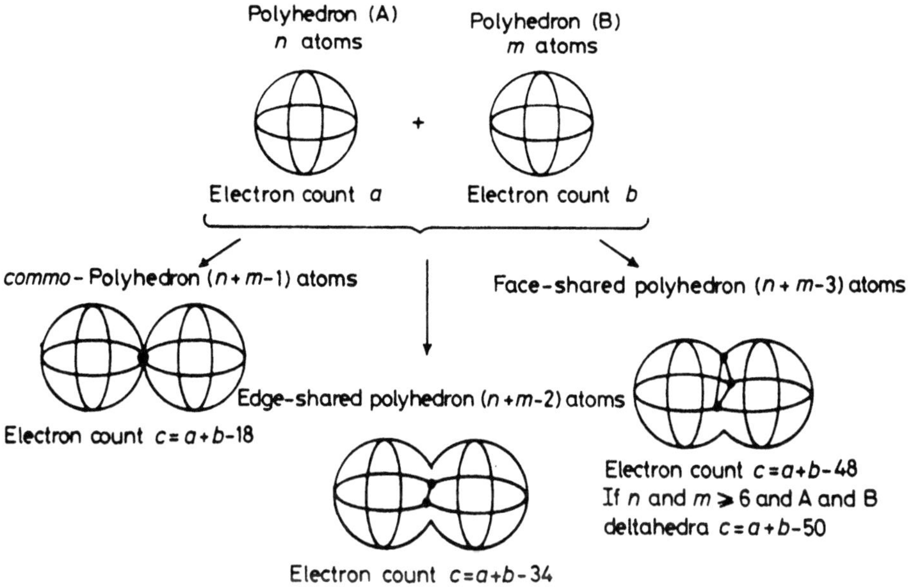

Fig. 11. Schematic representation of the principle of Polyhedral Fusion

polyhedron (B) by b electrons the condensed cluster (C) has a total of ($a + b - \Delta_c$) electrons (see Fig. 11). Table 5 lists the characteristic electron counts (Δ_c) for various shared units, based on the 18 electron rule, and example of this analysis of condensed clusters are presented in Fig. 12. Mingos and Evans have subsequently extended the Polyhedral Fusion approach to include condensed platinum clusters where the parent clusters and shared units are characterised by fewer electrons than in the previously mentioned examples (16 electrons for a single Pt atom, 30 electrons for a dimer, 42 or 44 electrons for a triangle etc.)[120]. The methodology has also been applied to condensed mixed main group-transition metal clusters[119].

Table 5. Characteristic electron (Δ_c) for shared units in fused transition metal polyhedra

Shared unit	Δ_c	Molecular analogue
Vertex	18	$[Mo(CO)_6]$
Edge	34	$[Mn_2(CO)_{10}]$
Triangular Face	48[a]	$[Os_3(CO)_{12}]$
	50[b]	–
Square Face	62[b]	$[Ru_4(CO)_{11}(\mu^4\text{-}PPh)_2]$
	64[a]	$[Fe_4(CO)_{12}(\mu^4\text{-}PPh)_2]$
Butterfly	62	$[Os_4N(CO)_{12}]^-$

[a] These values are used when the fused polyhedra are both 3-connected

[b] These values are used when one, or both, of the fused polyhedra are not 3-connected (Mingos, D. M. P.: Acc. Chem. Res. *17*, 311 (1984).

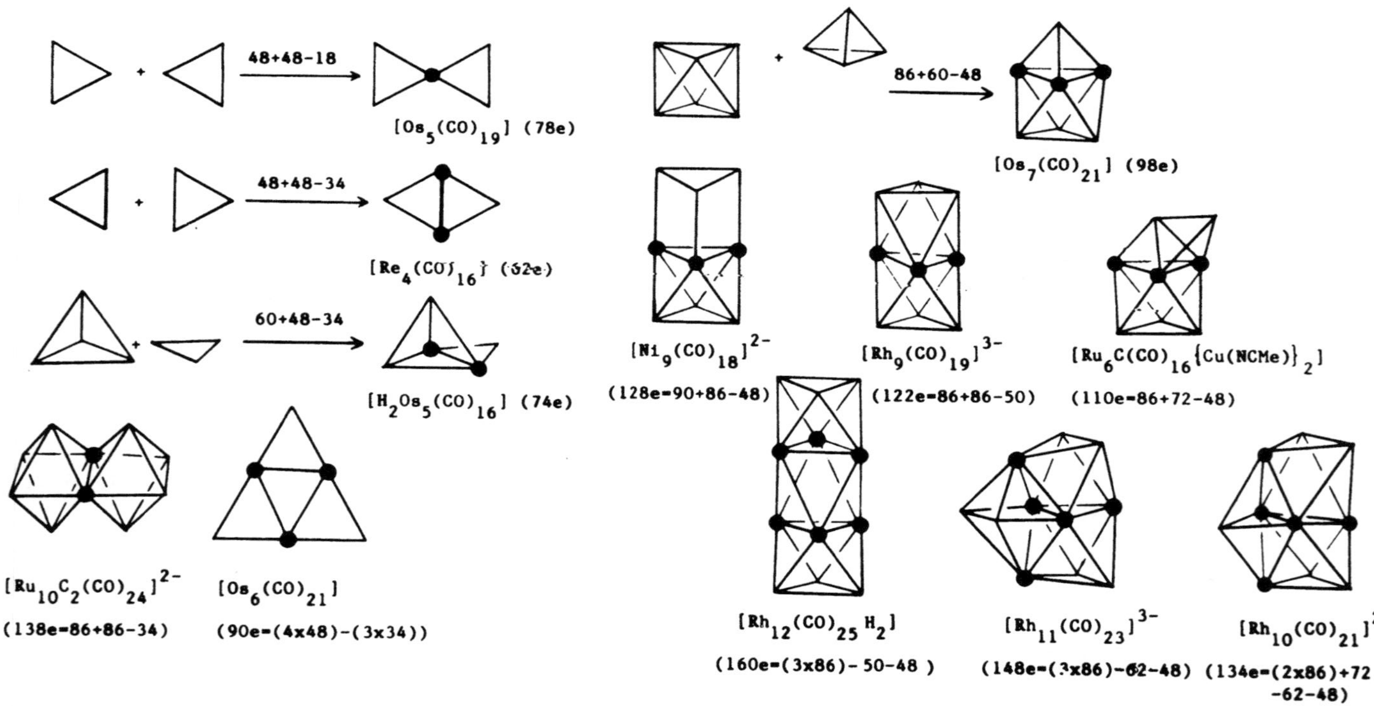

Fig. 12. Examples of the application of the principle of Polyhedral Fusion

2. *The Analogy Between High Nuclearity Clusters and Bulk Metals*

In the 1970's, the idea that discrete transition metal clusters could serve as good models for chemisorption onto bulk metal surfaces was elegantly formulated by Muetterties[8)]. Subsequently, a large number of high nuclearity clusters, with structures resembling fragments of bulk metal lattices (i.e. body centred cubic (b.c.c.), cubic close packed (c.c.p.) and hexagonal close packed (h.c.p.)), have been characterised, notably by Chini and coworkers[121)].

Lauher (in 1979) performed Extended Hückel calculations on a series of bare metal close packed clusters, partitioning the cluster valence MO's into those involved in metal-metal bonding and those which can act as acceptor orbitals towards ligands or substrates bound to the surface of a metal[122)]. In 1980, Ciani and Sironi published an Extended Hückel analysis of a wider selection of bare metal condensed clusters with known and hypothetical skeletal geometries[123)]. They concluded, on the bais of Lauher's criteria, that the majority of these clusters are characterised by (6 n + 7) cluster valence MO's. This has been rationalised in terms of the Capping Principle by recognising that most of these structures are based on a central octahedron which possesses 7 SEP's[124)].

Recently, Teo has used an extended form of the Hume-Rothery rule (developed for intermetallic alloys[125)] to account for the structure adopted by pseudo-close packed clusters[126)]. His conclusions are in agreement with the earlier work of Johnson, Messmer et al. who found that the density of states of a bare metal cluster rapidly converges to that of the bulk metal for as few as 13 (f.c.c.) or 15 (b.c.c.) metal atoms[127)].

3. *The Principle of Polyhedral Inclusion*

Mingos has recently developed an electron counting procedure (which may be described as ***"Polyhedral Inclusion"***) which is based on the formal division of the high nuclearity cluster into an internal (encapsulated) polyhedron and an external (surface) polyhedron[128)]. Three sub-classes of "close-packed" cluster have been identified:

Table 6. Characteristic electron counts (Δ_i) for interstitial moieties in high nuclearity clusters

No. of interstitial atoms	Δ_i	Molecular analogue
1	18	$[Mo(CO)_6]$
2 (dimer)	34	$[Mn_2(CO)_{10}]$
3 (triangle)	48	$[Os_3(CO)_{12}]$
	50[a]	–
3 (linear fragment)	50	$[OsRe_2(CO)_{14}]$
4 (tetrahedron)	60	$[Ir_4(CO)_{12}]$
6 (octahedron)	86	$[Rh_6(CO)_{16}]$

[a] Although isolated triangular clusters are characterised by 48 electrons, the presence of bridging metal atoms can lead to the stabilisation of an a_2' molecular orbital and a valence electron count of 50 for the central triangle (Evans, D. G., Mingos, D. M. P.: Organometallics *2*, 435 (1983).

Table 7. Examples of high nuclearity cluster compounds where radial metal-metal bonding interactions predominate

Cluster	Ref.	n_i	n_s	Structure	Electron count	
$[Au_9(PPh_3)_8]^+$	(a)	1	8	b.c.c.	114	
$[Au_{11}I_3(PPh_3)_7]$	(b)	1	10	b.c.c./i.c.p.[j]	138	$(12\,n_s + 18)$
$[Au_{13}Cl_2(PMePh_2)_{10}]^{3+}$	(c)	1	12	i.c.p.	162	
$[Pt_{19}(CO)_{22}]^{4-}$	(d)	2	17	i.c.p.	238	$(12\,n_s + 34)$
$[Rh_{22}(CO)_{35}H_{(5-q-m)}]^{q-}$	(e)	2	20	c.c.p./b.c.c.	$(237 + m)$[k]	
$[Au_{13}Ag_{12}Cl_6(PPh_3)_{12}]^{m+}$	(f)	3	22	i.c.p.	$(317 - m)$	$(12\,n_s + 50)$[l]
$[Pt_{26}(CO)_{32}H_x]$	(g)	3	23	h.c.p.	$324 + x$	$(12\,n_s + 48 + x)$[m]
$[Ni_{38}Pt_6(CO)_{48}H_{(6-n)}]^{n-}$	(h)	6	38	c.c.p.	542	$(12\,n_s + 50)$[n]
$[Pt_{38}(CO)_{44}H_m]^{2-}$	(i)	6	32	c.c.p.	$(470 + m)$[o]	

(a) van der Linden, J. G. M., Paulissen, M. L. H., Schmitz, J. E. J.: J. Amer. Chem. Soc. *105,* 1903 (1983).
(b) Smits, J. M. M., Burskens, P. T., van der Velden, J. W. A., Bau, J. J.: J. Cryst. Spectrosc. Res. *13,* 373 (1983).
(c) Briant, C. E., Theobald, B. R. C., White, J. W., Bell, L. K., Mingos, D. M. P.: J. Chem. Soc, Chem. Commun. 201 (1981).
(d) Waschecheck, D. M., Wucherer, E. J., Dahl, L. F., Ceriotti, A., Longoni, G., Manassero, M., Sansoni, M., Chini, P.: J. Amer. Chem. Soc. *101,* 6110 (1979)
(e) Vidal, J. L., Schoening, R. C., Troup, J. M.: Inorg. Chem. *20,* 227 (1981).
(f) Teo, B. K., Keating, K.: J. Amer. Chem. Soc. *106,* 2224 (1984).
(g) Ceriotti, A., Chini, P., Longoni, G., Waschecheck, D. M., Wucherer, E. J., Dahl, L. F., Marchionna, M., Montag, R. A.: unpublished results.
(h) ($n = 4$ or 5) Ceriotti, A., Demartin, F., Longoni G., Manassero, M., Marchionna, M., Piva, G., Sansoni, M.: Angew. Chem. Int. Edn. Engl. *24,* 696 (1985).
(i) Waschecheck, D. M., Ceriotti, A., Murphy, M. A., Nagaki, D. A., Chini, P., Longoni, G., Dahl, L. F.: personal communication.
(j) i.c.p. = "icosahedral close packing", a term used to denote high nuclearity clusters with 5-fold symmetry (Mingos, D. M. P.: Chem. Soc. Rev. *15,* 31 (1986).
(k) If $m = 1$ this cluster possesses 274 $(12\,n_s + 34)$ valence electrons.
(l) Linear trimeric interstitial moiety. If $m = 3$ this cluster possesses 314 $(12\,n_s + 50)$ valence electrons.
(m) Triangular interstitial moiety.
(n) Octahedral interstitial moiety.
(o) If $m = 0$ this cluster possesses 470 $(12\,n_s + 86)$ valence electrons.

(a) Clusters where Radial Metal-Metal Bonding Predominates

In this case, the surface atoms act as ligands towards the central atom or cluster, to which they are bonded by radial (σ) bonds. The surface tangential bonding is not strong in these clusters. Cluster valence electron counts (N_e) for such clusters are given by:

$$N_e = 12\,n_s + \Delta_i \tag{8}$$

where n_s is the number of surface atoms and Δ_i is the electron count characteristic of the central atom or cluster (see Table 6). Examples of this class of cluster, which includes the centred gold clusters mentioned above (where a $(12\,n_s + 16)$ electron count is possible for toroidal topologies), are listed in Table 7 (taken from Ref. 128b) and some structures are depicted in Fig. 13.

An important feature of these clusters is that their structures are independent of the electron count of the surface polyhedron. Since radial metal-metal bonding predominates, there is a soft potential energy surface for cluster rearrangement. This has been experimentally observed for gold clusters[129)]. Theoretical studies of bulk metals have indicated that alternative close packed (and related) structures are also separated by small energy differences[130)].

(b) Clusters with a Partial Bonding Contribution from the Surface Tangential Orbitals

All of the clusters of this type (see Table 8 and Fig. 14 for examples) possess a single interstitial atom and are characterised by ($12\ n_s + 18 + x$) cluster valence electrons. These clusters generally have 3 extra occupied skeletal MO's (i.e. $x = 6$), which are exclusively involved in tangential bonding in the external cluster polyhedron.

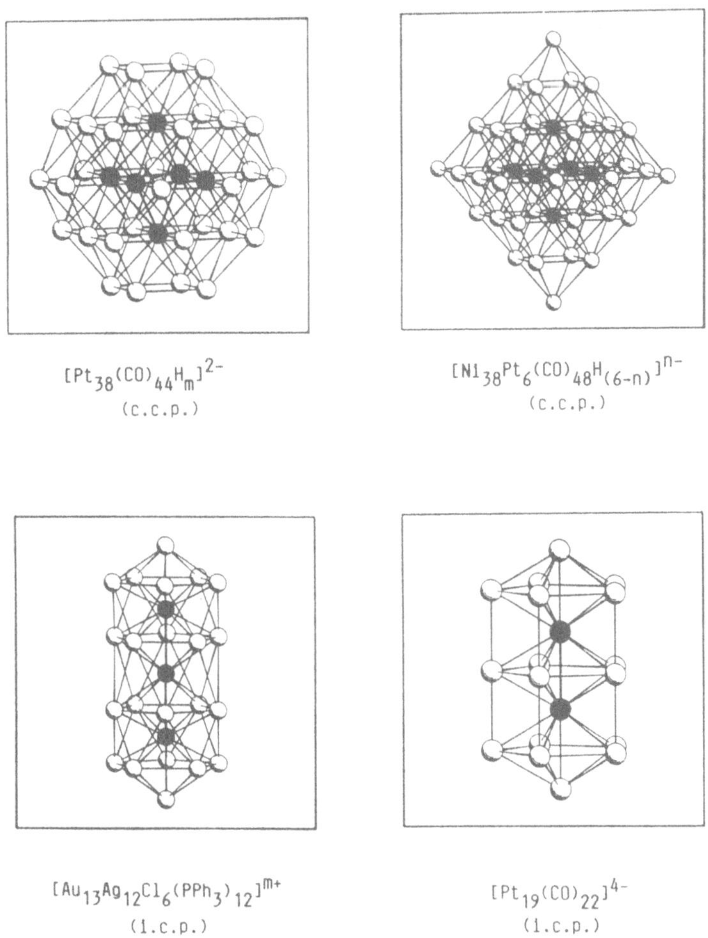

Fig. 13. Examples of high nuclearity transition metal clusters, with interstitial metal atoms and cluster fragments, where radial metal-metal bonding predominates

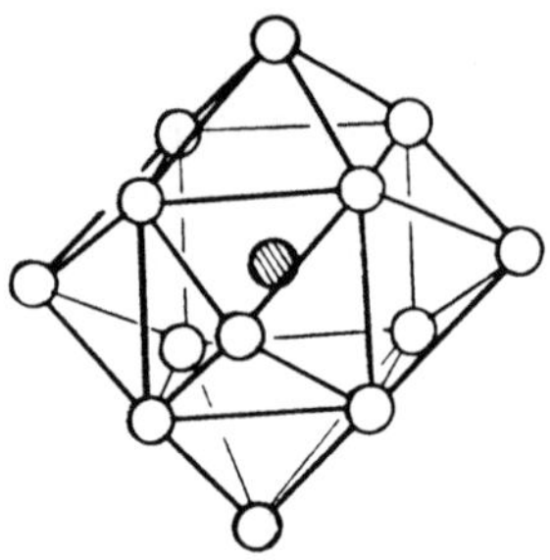

$[Rh_{14}(CO)_{26}]^{2-}$ (b.c.c.)

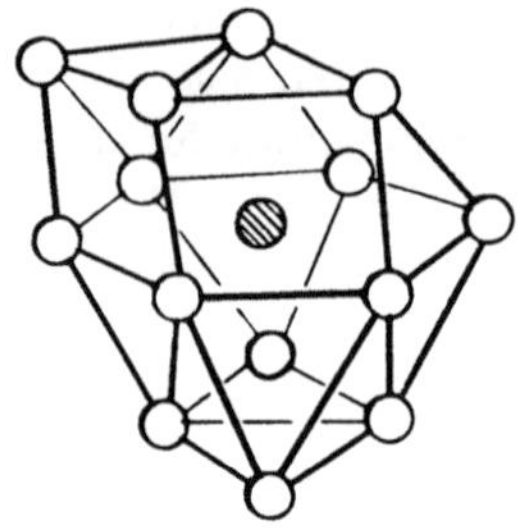

$[Rh_{15}(CO)_{27}]^{3-}$ (b.c.c./h.c.p.)

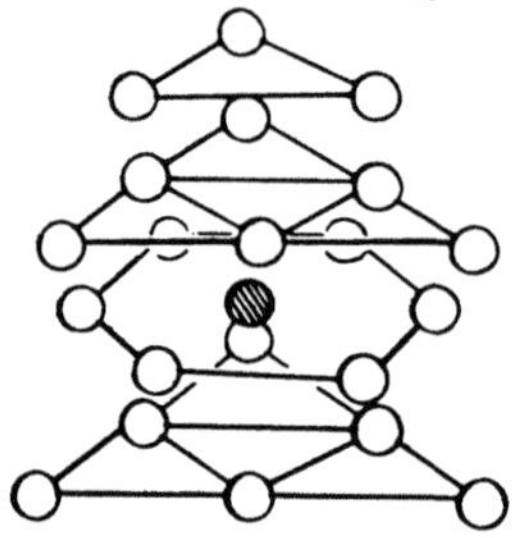

$[Rh_{22}(CO)_{37}]^{4-}$ (h.c.p./c.c.p.)

Fig. 14. Examples of high nuclearity transition metal clusters, with interstitial metal atoms, where the surface tangential orbitals make a partial contribution to the metal-metal bonding

Tabel 8. Examples of high nuclearity cluster compounds where tangential interactions make a partial contribution to metal-metal bonding

Cluster	Ref.	n_i	n_s	Structure	Electron count	
$[Rh_{14}(CO)_{25}]^{4-}$	(a)	1	13	b.c.c.	180	
$[Rh_{14}(CO)_{26}]^{2-}$	(b)	1	13	b.c.c.	180	
$[Rh_{15}(CO)_{27}]^{3-}$	(c)	1	14	b.c.c./h.c.p.	192	$(12\ n_s + 24)$
$[Rh_{17}(CO)_{30}]^{3-}$	(d)	1	16	h.c.p.	216	
$[Rh_{22}(CO)_{37}]^{4-}$	(e)	1	21	c.c.p./h.c.p.	276	
$[Pt_{24}(CO)_{30}]^{2-}$	(f)	1	23	c.c.p.	302	$(12\ n_s + 26)$

(a) Ciani, G., Sironi, A., Martinengo, S.: J. Organomet. Chem. *192,* C42 (1980).
(b) Martinengo, S., Ciani, G., Sironi, A.: J. Chem. Soc. Chem. Commun. 1140 (1980).
(c) Martinengo, S., Ciani, G., Sironi, A., Chini, P.: J. Amer. Chem. Soc. *100,* 7096 (1978).
(d) Ciani, G., Magni, A., Sironi, A., Martinengo, S.: J. Chem. Soc., Chem. Commun. 1280 (1981).
(e) Martinengo, S., Ciani, G., Sironi, A.: J. Amer. Chem. Soc. *102,* 7564 (1980).
(f) Montag, R. A., Ceriotti, A., Dahl, L. F.: personal communication.

(c) Clusters with a Complete Bonding Contribution from the Surface Tangential Orbitals

The electron count in these clusters (shown in Fig. 15), as for main group clusters, is governed by the arrangement of the skeletal (surface) atoms. For *closo*-type structures the electron count is $(14\ n_s + 2)$.

4. Physical Properties of High Nuclearity Clusters

Woolley has compared the metal-metal bonding in transition metal clusters and bulk metals and concluded that the metal d orbitals contribute significantly to this bonding in both cases[98, 99]. He has noted, however, that the presence of ligands in transition metal clusters modified the bonding properties of the d-band. The idea that the ligands of a cluster effectively insulate the metal atoms has been proposed by Mingos to explain why clusters have so far proved disappointing as regards their catalytic properties[131]. It is, however, still possible that theoretical studies of metal clusters will lead to an increased understanding of reactions on metal surfaces[132]. In addition, the study of large metal cluster compounds may lead to advances in the field of Materials Science, since interesting magnetic properties have recently been reported for a number of high nuclearity clusters[133]. These properties have led Benfield to postulate the existence of a new "metametallic" state[134].

D. Free Electron Cluster Bonding Models

I. Early Studies

As early as 1949, Pimental and Pitzer suggested[135] that the skeletal bonding in *nido*-decaborane $[B_{10}H_{14}]$ could be conveniently described in terms of a free electron model similar to those used for bulk metals[136]. Such a method had previously been applied to

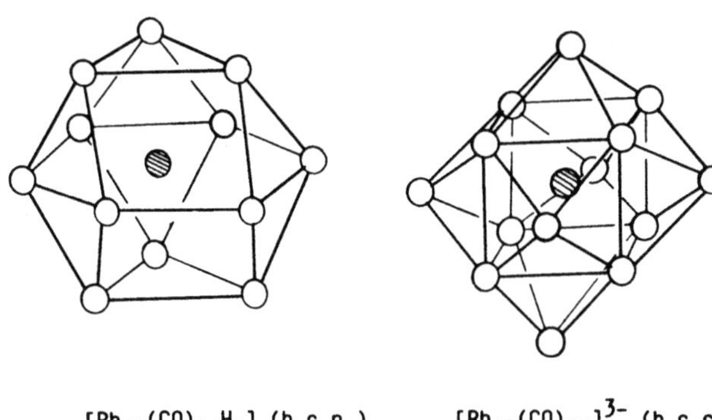

$[Rh_{13}(CO)_{24}H_5]$ (h.c.p.) $[Rh_{15}(CO)_{30}]^{3-}$ (b.c.c.)

Fig. 15. Examples of high nuclearity transition metal clusters, with interstitial metal atoms, where the surface tangential orbitals make a complete contribution to the metal-metal bonding the study of conjugated polyenes and aromatic hydrocarbones[137]. Subsequently, Chapman and Waddington treated the molecular orbitals of the cage compounds [S_4N_4] and [S_4As_4][138] in terms of the solution of the Schrödinger Equation for a particle on a sphere[22] (i.e. by allowing full delocalisation of the cluster valence electrons over the surface of a sphere)[139]. The cluster molecular orbitals were described, as for atomic orbitals, in terms of the quantum numbers L and M (with L giving the number of angular nodes possessed by the orbital and $M = L, L - 1, \ldots 0, \ldots -L$)[22]. In this simple analysis the $(2L + 1)$ components of each L manifold (i.e. orbitals with different M values) were taken to be degenerate, with their energies (E_L) given by the following equation:

$$E_L = \hbar^2 L(L + 1)/8\pi^2\mu a^2) \quad (9)$$

where a is the radius of the cluster sphere. The actual symmetry of the cluster (since the skeletal atoms must define a polyhedron with less than spherical symmetry) was then introduced as a perturbation, thereby breaking the $(2L + 1)$-fold degeneracy of the L manifold. Hoffmann and Gouterman used a similar mode of analysis[140], but approximated the perturbation (due to the real cluster geometry) by a Crystal Field[141] expansion. Waddington subsequently adapted his earlier model by constraining the skeletal electrons, in boranes and related species, to move along the edges of the cluster polyhedron (i.e. solving the Schrödinger Equation for a particle on the edges of a box). His methodology involved a topological factorisation of the secular equation[142].

These free electron models were successfully applied to linear and planar hydrocarbon π-systems and spherical aggregates of hydrogen atoms because, in these case, the nodal characteristics of the molecular orbitals accurately reflect their relative energies. For clusters of real interest it was found, however, that the energy level orderings obtained in this fashion were often different to those obtained by LCAO-MO methods[140]. This is because real clusters possess, in addition to nodeless (σ-type) atomic orbitals, orbitals which are singly (π) or doubly noded (δ) with respect to a plane perpendicular to a radial vector passing through the atom, as shown in Fig. 16. For linear and planar hydrocarbons, the π-system may be regarded as being topologically equivalent to an arrangement of hydrogen atoms, since the constituent p_z atomic orbitals are nodeless when viewed along an axis perpendicular to the plane of the molecule. Node counting for (3-dimensional) cluster MO's composed of atomic π and δ orbitals is complicated by the fact that these atomic orbitals possess intrinsic nodes which do not contribute to the overall antibonding character of the cluster MO's. Clearly a more sophisticated approach was needed to deal with π and δ as well as σ-type cluster MO's.

The creation of a more sophisticated spherical model for clusters was aided by the work of Hoffman, Ruedenberg and Verkade who developed a pictorial approach to the

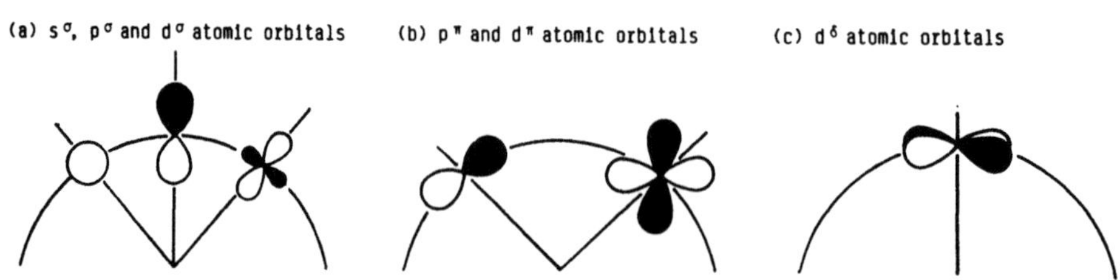

Fig. 16. σ-, π- and δ-symmetry atomic orbitals used in the construction of cluster MO's

bonding in main group and transition metal complexes[143]. They used Spherical Harmonic functions, at the coordinate origin of the molecule, as Generator Orbitals to create linear combinations of ligand orbitals with the same symmetry and nodal characteristics[143]. These "symmetry orbitals" were then combined with the orbitals of the real central atom of the molecule in order to form localised or delocalised orbitals. The generator orbital was identified with the angular part of a united atom orbital (a concept introduced by Mulliken[144], where the electrons of a molecule were regarded as being delocalised within a sphere rather than over its surface.

II. Tensor Surface Harmonic Theory

1. Introduction: Mathematical Aspects and Examples

Recently, an elegant approach to the bonding in cluster compounds has been developed by A. J. Stone, whose ***Tensor Surface Harmonic (TSH) Theory***[145] derives cluster molecular orbitals from a spherical potential model. All of the cluster atoms are constrained to lie on the surface of a single sphere, of unit radius, with each atom (i) being described by the angular coordinates (θ_i, ϕ_i). This is followed by evaluation of the angular parts of the solutions of the Schrödinger equation for a particle on a sphere[22] (the radial parts can be ignored if the cluster vertices are taken to lie on the surface of a sphere of unit radius)[145a]:

$$\nabla^2\Psi = -L(L+1)\Psi \tag{10}$$

where $\nabla^2 = (\sin\theta)^{-1}(\partial/\partial\theta)\sin\theta(\partial/\partial\theta) + (\sin^2\theta)^{-1}(\partial^2/\partial\phi^2)$. This yields wavefunctions which are Spherical Harmonic[146] in nature and may (as in the treatment of Chapman and Waddington[139]) be described in terms of the integer quantum numbers L and M($L > 0$; $M = L, L-1 \ldots -L$). Stone's method differs from the earlier free electron approaches in that the values of the general Spherical Harmonics $S_{L,M}(\theta, \phi)$ at the cluster atom positions (θ_i, ϕ_i) are used as the coefficients (c_i) in an LCAO expansion of a bais set of atomic orbitals (ϱ_i) to yield the cluster MO's ($\Psi_{L,M}$):

$$\Psi_{L,M} = N \sum_i c_i\varrho_i = N \sum_i S_{L,M}(\theta_i, \phi_i)\varrho_i \tag{11}$$

The next stage in the TSH analysis depends upon whether the cluster MO's to be generated are based on atomic orbitals with 0(σ), 1(π) or 2(δ) nodes with respect to a radial vector passing through the atoms (see Fig. 16).

(a) Radial σ-Type Orbitals

For a basis set of atomic (σ) orbitals which are nodeless in the plane perpendicular to the radial vector (see Fig. 16a), the required Harmonics are simply the Scalar Spherical Harmonics ($Y_{L,M}(\theta, \phi)$)[146]. The LCAO expansion, therefore, takes the following form:

$$\Psi^{\sigma}_{L,M} = \sum_i Y_{L,M}(\theta_i, \phi_i)\sigma_i \tag{12}$$

LCAO coefficients are required to be real numbers, so (for M > 0) the real form of the Spherical Harmonics are required:

$$Y_{L,Mc} = \sqrt{2}\{(-1)^M Y_{L,M} + Y_{L,-M}\}$$
$$Y_{L,Ms} = \sqrt{2}i\{(-1)^M Y_{L,M} - Y_{L,-M}\} \quad (13)$$

The subscripts "Mc" and "Ms" refer to the cosine and sine dependence on ϕ which these functions possess:

$$Y_{L,Mc} = \sqrt{2}\, Y_{L,M}(\theta)\cos(M\phi)$$
$$Y_{L,Ms} = \sqrt{2}\, Y_{L,M}(\theta)\sin(M\phi) \quad (14)$$

(this follows from the separation of $Y_{L,M}(\theta, \phi)$ into θ and ϕ components, i.e. $Y_{L,M}(\theta, \phi) = Y_{L,M}(\theta)e^{iM\phi}$).

The σ-type cluster wavefunctions ($\Psi_{L,Mc}/\Psi_{L,Ms}$) are generally written as $L^{\sigma}_{Mc}/L^{\sigma}_{Ms}$, where the orbitals are labelled S(L = 0), P(L = 1), D(L = 2), F(L = 3) etc. For an n-vertex clusters, the first nL^{σ} functions (i.e. those with the lowest L and M values) are chosen, provided that:
(i) they do not possess zero amplitude at all of the cluster vertices and
(ii) they are not merely repeats of earlier MO's[145a)].

Examples of the L^{σ} MO's of some small clusters are depicted in Fig. 17. The energies of the L^{σ}_{M} cluster MO's ($E^{\sigma}_{L,M}$) are obtained, via the Variation Principle[147)], as:

$$E^{\sigma}_{L,M} = \frac{\int \Psi^{\sigma}_{L,M}{}^{*}\hat{\mathscr{H}}\Psi^{\sigma}_{L,M}\,d\tau}{\int \Psi^{\sigma}_{L,M}{}^{*}\Psi^{\sigma}_{L,M}\,d\tau} \quad (15)$$

where $\hat{\mathscr{H}}$ is the effective one electron Hamiltonian of the system. In the original TSH theory, the assumption that the energies of the cluster L^{σ} MO's are reasonably independent of M was made. This leads to a (2L + 1)-fold degeneracy for each L^{σ} manifold. In many cases, however, not all of the (2L + 1) components of a given L^{σ} manifold are valid solutions for the cluster. This means that the average energy (E^{σ}_{L}) of the L^{σ} sub-shell cannot be obtained from the simple average ($\Sigma_M E^{\sigma}_{L,M}/(2L + 1)$), but instead has to be obtained from the following weighted mean[145a)]:

$$E^{\sigma}_{L} = \frac{\sum_{M} \int \Psi^{\sigma}_{L,M}{}^{*}\hat{\mathscr{H}}\Psi^{\sigma}_{L,M}\,d\tau}{\sum_{M} \int \Psi^{\sigma}_{L,M}{}^{*}\Psi^{\sigma}_{L,M}\,d\tau} \quad (16)$$

This equation may alternatively be written in terms of Spherical Harmonics:

$$E^{\sigma}_{L} = \frac{\sum_{ij}\sum_{M} Y_{L,M}(\theta_i, \phi_i)^{*}Y_{L,M}(\theta_j, \phi_j)H_{ij}}{\sum_{ij}\sum_{M} Y_{L,M}(\theta_i, \phi_i)^{*}Y_{L,M}(\theta_j, \phi_j)S_{ij}} \quad (17)$$

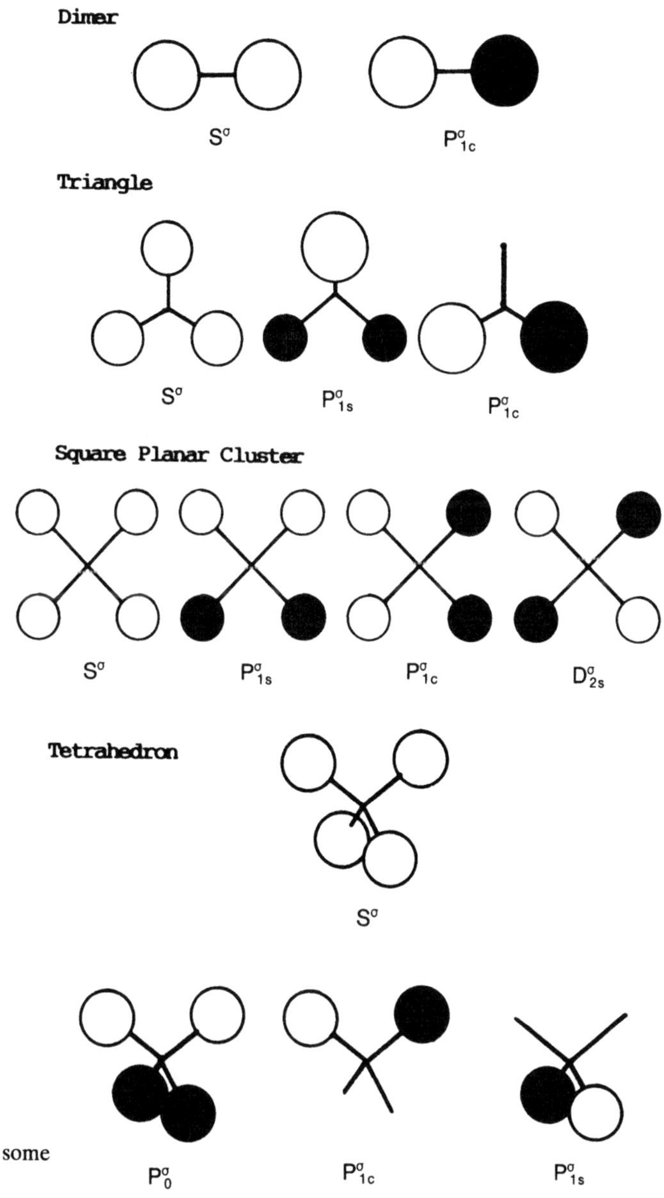

Fig. 17. The L^{σ} MO's of some low nuclearity clusters

where $H_{ij} = \int \sigma_i \mathscr{H} \sigma_j \, d\tau$ and S_{ij} (Overlap Integral) $= \int \sigma_i \sigma_j \, d\tau$. By applying the Spherical Harmonic Addition Theorem[146)], the numerator of the above equation may be written in terms of the angular separation (ω_{ij}) between each pair of atoms, i and j:

$$= \sum_{ij} [(2L + 1)/4\pi] P_L(\cos \omega_{ij}) H_{ij} \tag{18}$$

where $P_L(\cos \omega_{ij})$ is a Legendre Polynomial of the form:

$P_L(x) = (2^L L!)^{-1} d^L/dx^L (x^2 - 1)^L$

The next step is to make the simple Hückel approximations[148]: $S_{ij} = \delta_{ij}$; $H_{ij} = \alpha^\sigma(i = j)$, $H_{ij} = \beta^\sigma$(i, j adjacent) or else $H_{ij} = 0$. The analysis is further simplified by making the approximation that the angular separations of all adjacent atoms are equal (ω). This enables the following equation to be derived[145a,b].

$$E^\sigma_L = \alpha^\sigma + (2e/n)\beta^\sigma P_L(\cos\omega) \quad (19)$$

where e is the number of edges of the cluster polyhedron (i.e. number of pairs of adjacent atoms) and n is the number of cluster vertices. The energy level scheme for these orbitals is such that the energy increases with the number of angular nodes (i.e. L) of the function, $S^\sigma < P^\sigma < D^\sigma$ etc.[145]. The factor (2e/n) is merely the average coordination number of the cluster atoms. This number is a maximum for deltahedral structures, which led to the conclusion that (since α^σ and β^σ are both negative) ***if only skeletal bonding molecular orbitals are occupied, the most stable cluster structures are the deltahedra.*** In this way Stone rationalised the predominance of such triangulated structures in cluster chemistry[145a,b].

(b) Surface π-Type Orbitals

For a basis set of (π) atomic orbitals which are singly noded in the plane perpendicular to the radial vector (see Fig. 16b), the required Harmonics are the Vector Surface Harmonics[146]. The two p^π (or d^π) atomic orbitals at each cluster vertex (i) behave as a pair of orthogonal unit vectors which are tangential to the surface of the sphere. π_i^θ and π_i^ϕ are defined as π-symmetry orbitals on the ith cluster atom, pointing in the direction of increasing θ and ϕ respectively, as shown in Fig. 18. At the poles of the cluster sphere these vectors may be related to Cartesian vectors as follows:

North Pole: $(\pi_i^\theta, \pi_i^\phi) = (\pi_i^x, \pi_i^y)$
South Pole: $(\pi_i^\theta, \pi_i^\phi) = (-\pi_i^x, \pi_i^y)$

Two Vector Surface Harmonic functions may be obtained from each (Scalar) Spherical Harmonic, one by taking its gradient ($\mathbf{V}_{L,M} = \nabla Y_{L,M}$) and the other by forming a

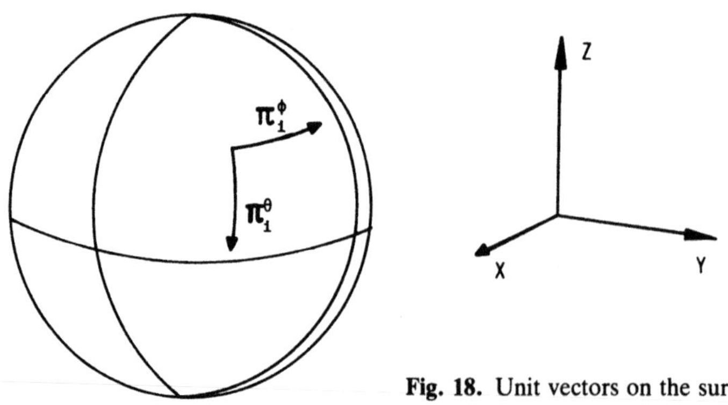

Fig. 18. Unit vectors on the surface of the cluster sphere

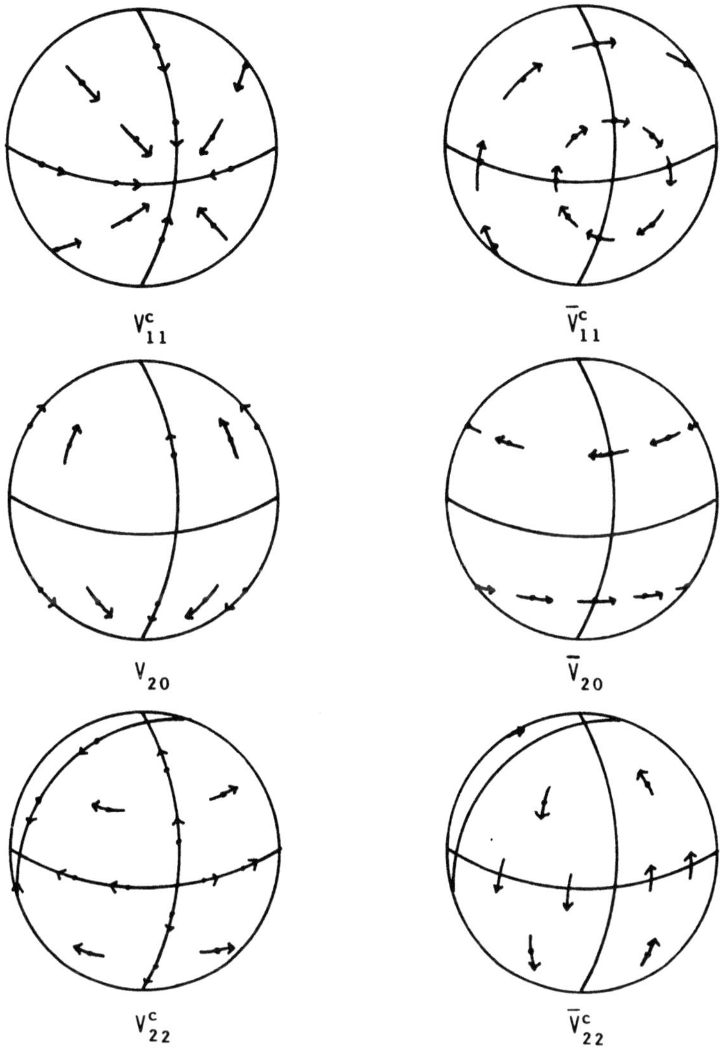

Fig. 19. Examples of even and odd parity Vector Surface Harmonic functions

vector function which is orthogonal to the first ($\bar{\mathbf{V}}_{L,M} = \hat{\mathbf{r}} \times \mathbf{V}_{L,M}$), where $\nabla = (\partial/\partial\theta) + (\sin\theta)^{-1}(\partial/\partial\phi)$. These functions can be said to possess ***even*** or ***odd parity***, depending on their behaviour under the inversion operator, $\hat{\mathcal{J}}$[145]:

$$\hat{\mathcal{J}}\mathbf{V}_{L,M} = (-1)^L \mathbf{V}_{L,M} \quad \text{(even parity)}$$
$$\hat{\mathcal{J}}\bar{\mathbf{V}}_{L,M} = (-1)^{L+1} \bar{\mathbf{V}}_{L,M} \quad \text{(odd parity)} \qquad (20)$$

where the functions $\mathbf{V}_{L,M}$ possess the same (even) parity as the $Y_{L,M}$ functions from which they are derived. There is no S^π orbital, since S^σ has no θ or ϕ dependence and, therefore, has no gradient. Examples of even and odd parity Vector Surface Harmonic functions are depicted in Fig. 19.

The Vector Surface Harmonic functions may be divided into two orthogonal components which indicate their θ and ϕ dependence:

$$\mathbf{V}_{L,M} = (\mathbf{V}^{\theta}_{L,M}, \mathbf{V}^{\phi}_{L,M})$$
$$\overline{\mathbf{V}}_{L,M} = (\overline{\mathbf{V}}^{\theta}_{L,M}, \overline{\mathbf{V}}^{\phi}_{L,M}) \qquad (21)$$

where:

$$\mathbf{V}^{\theta}_{L,M} = \overline{\mathbf{V}}^{\phi}_{L,M} = (\partial/\partial\phi)Y_{L,M}$$
$$\mathbf{V}^{\phi}_{L,M} = -\overline{\mathbf{V}}^{\theta}_{L,M} = (\sin\theta)^{-1}(\partial/\partial\phi)\,Y_{L,M} \qquad (22)$$

The even and odd parity functions are related by a 90° anticlockwise rotation of the tangential vectors about the radial vector ($\hat{r}$), at each point on the sphere. This is known as the Parity Inversion Operation and is described by the operator $\hat{\mathcal{P}}$[145]:

$$\hat{\mathcal{P}}\mathbf{V}_{L,M} = \overline{\mathbf{V}}_{L,M}$$
$$\hat{\mathcal{P}}\overline{\mathbf{V}}_{L,M} = -\mathbf{V}_{L,M} \qquad (23)$$

Cluster π-type wavefunctions are generated by using these Vector Surface Harmonics as the coefficients in an LCAO expansion, with a basis set of π_i^{θ} and π_i^{θ} orbitals:

$$\Psi^{\pi}_{L,M} = L^{\pi}_{M} = N \sum_i (\mathbf{V}^{\theta}_{L,M}(\theta_i, \phi_i)\pi_i^{\pi} + \mathbf{V}^{\phi}_{L,M}(\theta_i, \phi_i)\pi_i^{\pi}) \qquad (24)$$

(for M > 0 the real forms of the Vector Surface Harmonics ($\mathbf{V}_{L,Ms}/\mathbf{V}_{L,Mc}$) are used). A similar expression may be derived for the $\overline{\Psi}^{\pi}_{L,M}(\overline{L}^{\pi}_{M})$ wavefunctions. As for the L^{σ} functions, the first n non-zero (and non-repeating) L^{π} and $\overline{L}^{\pi}$ functions are used. The L^{π} and $\overline{L}^{\pi}$ MO's of an octahedral cluster are depicted in Fig. 20.

For these L^{π} and $\overline{L}^{\pi}$ MO's, similar energy expression may be derived to those obtained for the L^{σ} orbitals[145a)]:

$$E^{\pi}_{L} = \alpha^{\pi} + (2e/n)f^{\pi}_{L}(\omega)$$
$$\overline{E}^{\pi}_{L} = \alpha^{\pi} + (2e/n)\overline{f}^{\pi}_{L}(\omega) \qquad (25)$$

where $f^{\pi}_{L}(\omega)$ and $\overline{f}^{\pi}_{L}(\omega)$ are functions of ω which take account of the fact that there are σ and π components to the overlap between tangential p^{π} orbitals (for cluster MO's composed of atomic d^{π} orbitals, δ overlap must also be included). Stone showed that for deltahedral clusters the L^{π} MO's are generally bonding. The $\overline{L}^{\pi}$ MO's are, therefore, generally antibonding, since the parity inversion operation leads to a reversal of the bonding properties of L^{π} and $\overline{L}^{\pi}$[145c)]:

$$L^{\pi}_{M}\ \text{(bonding)} \xleftrightarrow{\hat{\mathcal{P}}} \overline{L}^{\pi}_{M}\ \text{(antibonding)}$$

Stone used the methodology of TSH theory to derive the (n + 1) skeletal electron pair count for deltahedral boranes[145a, b)]. Each of the (n) BH fragments possesses one σ-type sp hybrid, yielding nL^{σ} cluster MO's in total, one of which (S^{σ}) is strongly bonding because it possesses no angular nodes. In addition each fragment has two π-type p orbitals which combine to form nL^{π} bonding and $n\bar{L}^{\pi}$ antibonding cluster orbitals. Mixing occurs between L^{σ} and L^{π} orbitals, with the same L values, resulting in one bonding combination and one antibonding combination for every matched pair of L^{σ}/L^{π} orbitals (remembering that there is no S^{π} orbital). In total, therefore, there are (n + 1) cluster

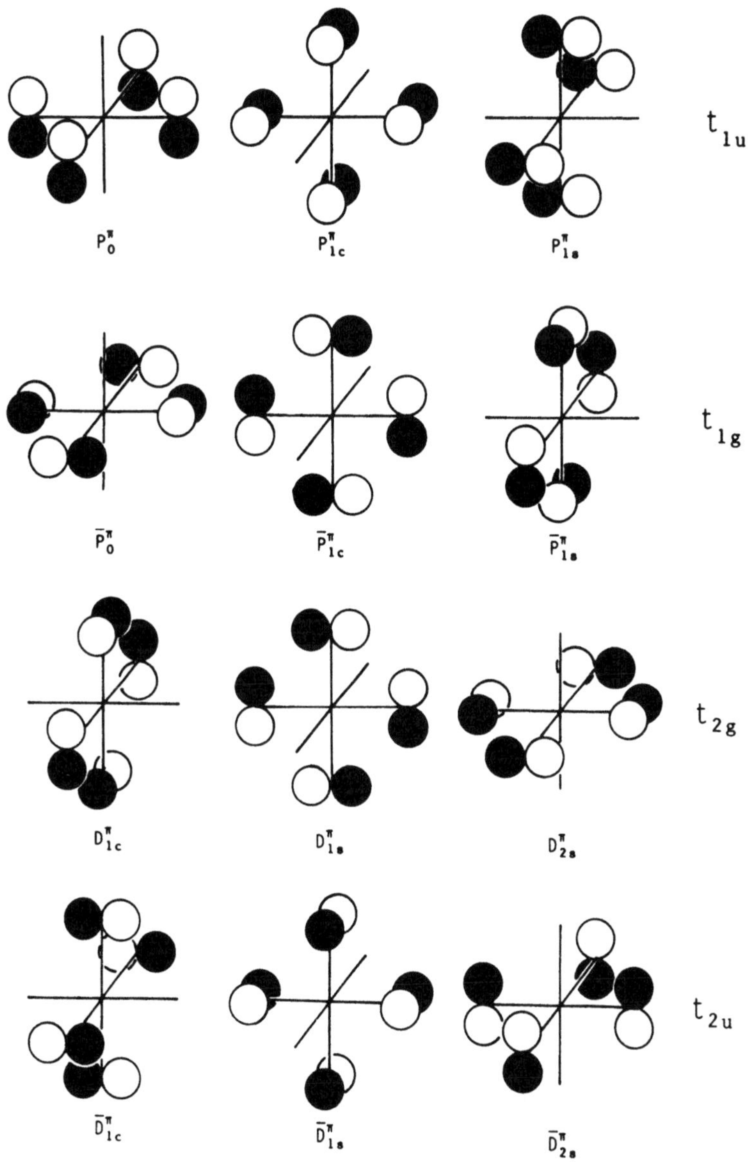

Fig. 20. The L^{π} and $\bar{L}^{\pi}$ MO's of an octahedral cluster

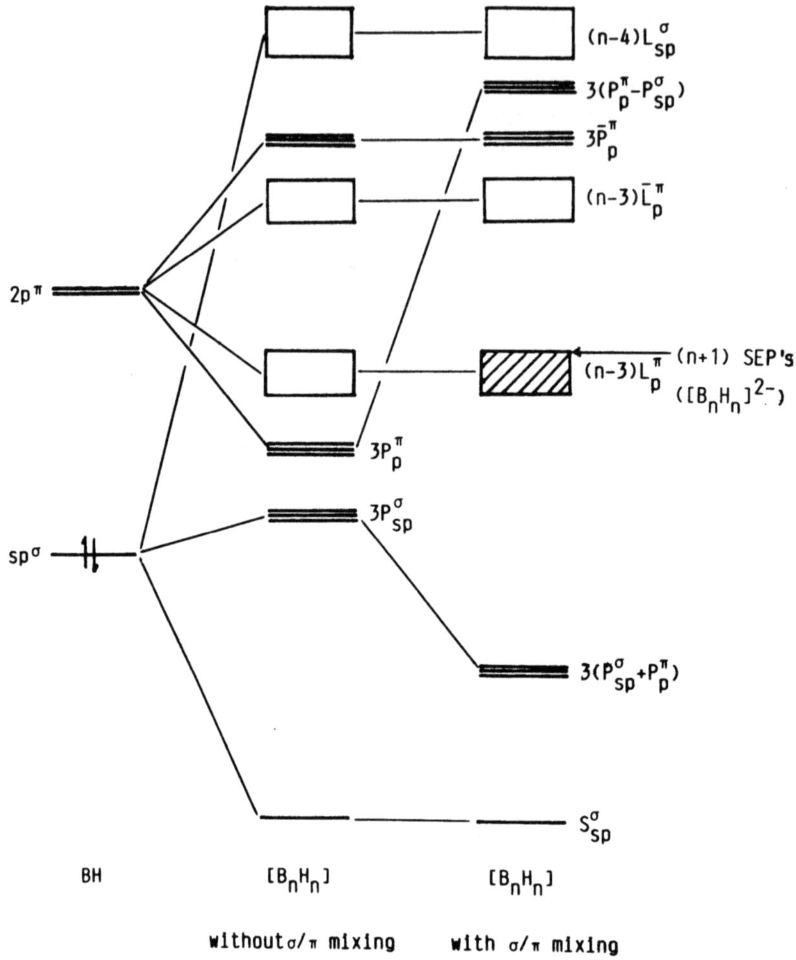

Fig. 21. Qualitative MO scheme, using Stone's TSH methodoloy, for the deltahedral borane anions $[B_nH_n]^{2-}$

bonding MO's (S^σ and nL^σ/L^π combinations). More sophisticated MNDO molecular orbital calculations have subsequently been performed by Brint et al., which have confirmed that the qualitative approach of TSH leads to a reasonably accurate description of the bonding in deltahedral borane clusters[149)]. They concluded that the only significant L^σ/L^π mixing occurs for the P^σ and P^π orbitals, since the higher L^σ orbitals are strongly antibonding and consequently, the L^σ-L^π energy separation is large. A qualitative molecular orbital scheme for the deltahedral boranes, based on TSH theory is presented in Fig. 21.

Stone has applied TSH theory to the study of *nido* and *arachno* borane clusters, deriving the (n + 2) and (n + 3) SEP counts for these species[145c)]. For instance, the *nido* clusters were found to possess a degenerate non-bonding pair of orbitals which are parity related ($L^\pi/\bar{L}^\pi$). The non-bonding nature of these orbitals may be rationalised in terms of the Parity Inversion Operator. This operator reverses the bonding characteristics of L^π

and $\overline{L}^{\pi}$ so for a degenerate $L^{\pi}/\overline{L}^{\pi}$ pair to exist, they must be non-bonding. The occupation of both of these orbitals results in an (n + 2) SEP count.

(c) Surface δ-Type Orbitals

To study the bonding in transition metal cluster compounds, a new type of Spherical Harmonic, with tensor properties, is required. This is because the metal d orbitals have (in addition to 1 σ and 2 π components) 2 δ components (d_{xy} and $d_{x^2-y^2}$) which are doubly noded in the plane perpendicular to the radial vector (see Fig. 16c) and which, therefore, behave as tensors. Two Tensor Surface Harmonic functions may be obtained from each Scalar Spherical Harmonic as follows[146]:

$$\begin{aligned} \mathbf{T}_{L,M} &= \nabla\nabla Y_{L,M} \text{ (even parity)} \\ \overline{\mathbf{T}}_{L,M} &= \hat{\mathbf{r}} \times \mathbf{T}_{L,M} \text{(odd parity)} \end{aligned} \tag{26}$$

The tensor Surface Harmonics are a measure of the concavity of the Scalar Spherical Harmonics. Since the concavities of S^{σ} and P^{σ} are zero, there are no S^{δ} or P^{δ} functions.

The Tensor Suface Harmonic functions may be expressed in the following matrix form:

$$\mathbf{T}_{L,M} = \begin{pmatrix} \mathbf{T}^{\theta\theta}_{L,M} & \mathbf{T}^{\theta\phi}_{L,M} \\ \mathbf{T}^{\phi\theta}_{L,M} & \mathbf{T}^{\phi\phi}_{L,M} \end{pmatrix} \qquad \begin{aligned} \overline{\mathbf{T}}^{\theta\theta}_{L,M} &= -\mathbf{T}^{\phi\phi}_{L,M} = -\mathbf{T}^{\theta\phi}_{L,M} \\ \overline{\mathbf{T}}^{\theta\phi}_{L,M} &= \overline{\mathbf{T}}^{\phi\theta}_{L,M} = \mathbf{T}^{\theta\theta}_{L,M} \end{aligned} \tag{27}$$

and may be used in an LCAO expansion, with a basis set of d^{δ} orbitals at each cluster vertex. These orbitals, $d_{\theta^2-\phi^2}$ and $d_{\theta\phi}$, are defined such that their positive lobes lie along the direction of increasing θ and the line θ = ϕ respectively. At the poles, they are equivalent to the Cartesian functions $d_{x^2-y^2}$ and d_{xy}.

In this way, the cluster wavefunctions $\Psi^{\delta}_{L,M}(L^{\delta}_{M})$ and $\Psi^{\delta}_{L,M}(\overline{L}^{\delta}_{M})$ are obtained. These wavefunctions are again related by the Parity Inversion Operation which, in this case, corresponds to a rotation of each d_{δ} component of the MO by 45° anticlockwise about a radial vector passing through the cluster vertex[145].

The energies of the L^{δ} and $\overline{L}^{\delta}$ orbitals are obtained, in an analogous method to that described above for the L^{σ} and $L^{\pi}/\overline{L}^{\pi}$ MO's, in terms of a function of ω, which takes into account the σ, π and δ components of the overlap between d^{δ} orbitals.

By employing the Scalar, Vector and Tensor Harmonics, Stone derived the (7 n + 1) cluster valence MO count for deltahedral transition metal clusters and later developed the model to provide an alternative derivation of Mingos' electron counting rules for condensed metal clusters[145d].

2. *Extended Tensor Surface Harmonic Theory*

The most serious drawback to Stone's elementary TSH theory is that all of the (2 L + 1) cluster MO's of a given L value are taken to be degenerate. For a real cluster, with non-spherical symmetry, the orbital degeneracies are generally lower than (2 L + 1). Fowler and Porterfield have circumvented this problem by introducing the Extended Hückel parameterisation of the Hamiltonian[65a,67] into the TSH framework, thus creating the ***Extended Tensor Surface Harmonic (ETSH) Theory***[150]. ETSH theory, which also takes quantitative account of σ-π mixing, leads to the correct orbital degeneracies under the actual point group symmetry of the cluster. The method has been applied to *closo* boranes in order to rationalise deviations from Wade's (n + 1) SEP rule[150]. The analysis of Fowler and Porterfield, together with the earlier studies of Brint et al.[149], has shown that the strength of TSH theory is that ***"its pseudo-spherical symmetry labels give a language for the qualitative discussion of cluster bonding that may still be useful in interpreting more quantitative calculations".***

3. Application of Tensor Surface Harmonic Theory to 3- and 4-Connected Clusters

In TSH theory, the solution of the Schrödinger Equation for the particle on a sphere problem generates L^{σ}, L^{π}, $\bar{L}^{\pi}$ etc. functions. For (hypothetical) truly spherical clusters the L^{π} and $\bar{L}^{\pi}$ functions constitute two distinct sets of orbitals which are strictly bonding and antibonding respectively. As indicated above, Stone has shown that the same bonding scheme may be applied to real cluster molecules which possess deltahedral geometries and belong to finite point groups. This is because a deltahedral arrangement of vertices is sufficiently spherical (i.e. represents a fairly good coverage of a spherical surface) for the cluster wavefunctions to represent good approximations to the particle on a sphere solutions.

We have used the TSH methodology to discuss the bonding in 4-connected polyhedral clusters (i.e. each cluster vertex has 4 near neighbours, see Fig. 22)[151]. These studies have led to the conclustion that 4-connected clusters, like deltahedral clusters, generally possess (n + 1) SEP's, with the L^{π} and $\bar{L}^{\pi}$ orbitals having bonding and antibonding character respectively. It can be seen, therefore, that 4-connected polyhedra also emulate spheres sufficiently well for the spherical limit of TSH theory to be applicable. If, however, the 4-connected polyhedron is distorted in such a way that its topology becomes prolate- or oblate-spheroidal (corresponding to capped 3-connected polyhedral geometries) then SEP counts other than (n + 1) may occur. This is due to the presence, in the frontier orbtal region, of approximately non-bonding parity-matched L^{π} and $\bar{L}^{\pi}$ orbitals. Figure 23 shows the correlation between the skeletal MO's of a 4-connected ("pseudo-spherical") cuboctahedral cluster and the ("oblate-spheroidal") tetracapped cube, to which it is related by a distortion. While the cuboctahedron possesses (n + 1 = 13) SEP's the tetracapped cube could possess n (12), (n + 1) or (n + 2) SEP's., due to the non-bonding parity-related $F_0^{\pi}(a_{2u})/\bar{F}_0^{\pi}(a_{2g})$ pair. The bonding and non-bonding MO's of 4-connected polyhedral clusters and related capped 3-connected polyhedra are listed in Table 9[151].

As mentioned in Sect. B, the bonding in 3-connected clusters, such as the hydrocarbon clusters $[C_nH_n]$ depicted in Fig. 2, may be described in terms of 3n/2 (2c − 2e) bonds which are localised along the cluster edges. We have applied TSH theory to study these molecules, in an attempt to rationalise bonding differences between 3-connected and deltahedral clusters[152].

The pattern of skeletal MO's characteristic of a general 3-connected cluster $[E_nH_n]$ (where E is a main group element) is contrasted in Fig. 24 with that for a general deltahedral cluster of the same nuclearity. The figure shows that, while deltahedra generally possess n bonding L^π and n antibonding $\bar{L}^\pi$ orbitals, 3-connected clusters are characterised by (n − 2) approximately non-bonding skeletal MO's, of which half are L^π and the other half are their parity-matched $\bar{L}^\pi$ counterparts. The occurence of non-bonding parity-related $L^\pi/\bar{L}^\pi$ sets is a consequence of the nature of the Parity Inversion Operation, as Stone explained for the *nido* boranes[145c)]. The bonding and non-bonding MO's (and

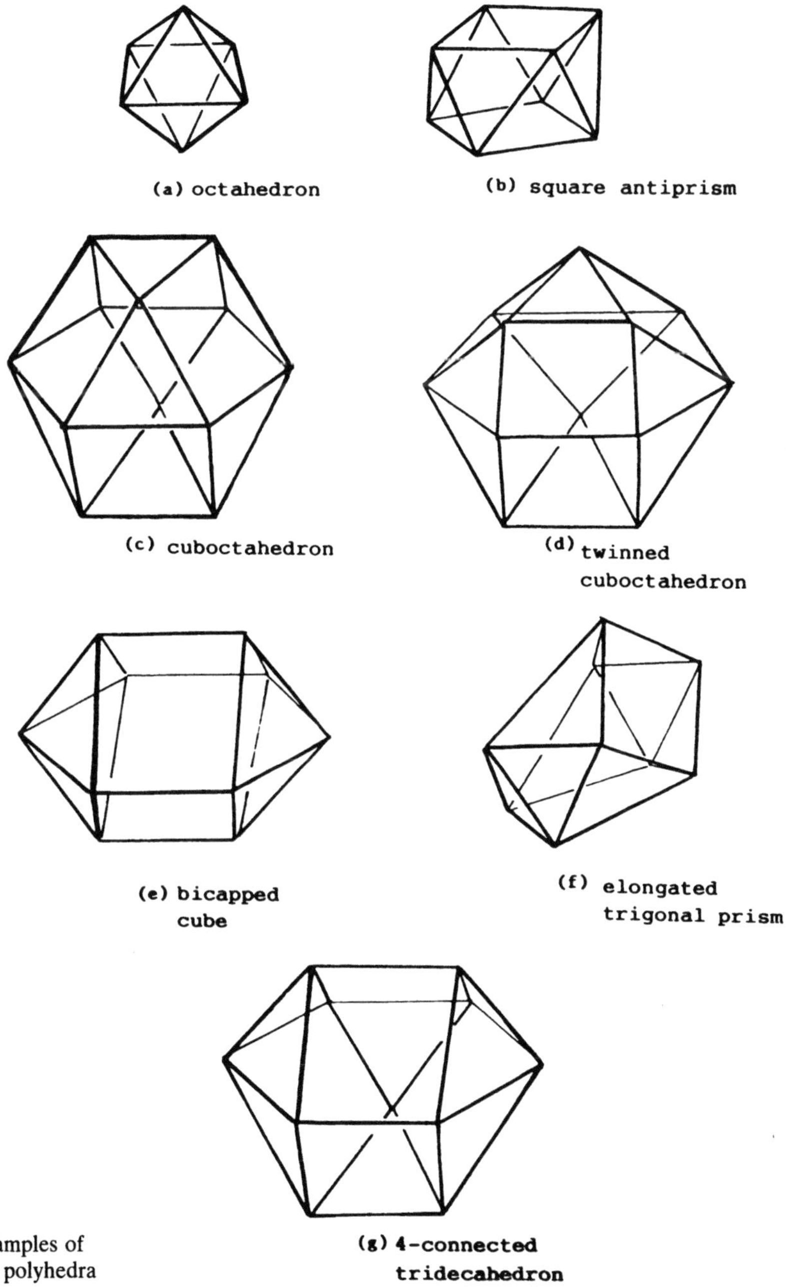

Fig. 22. Examples of 4-connected polyhedra

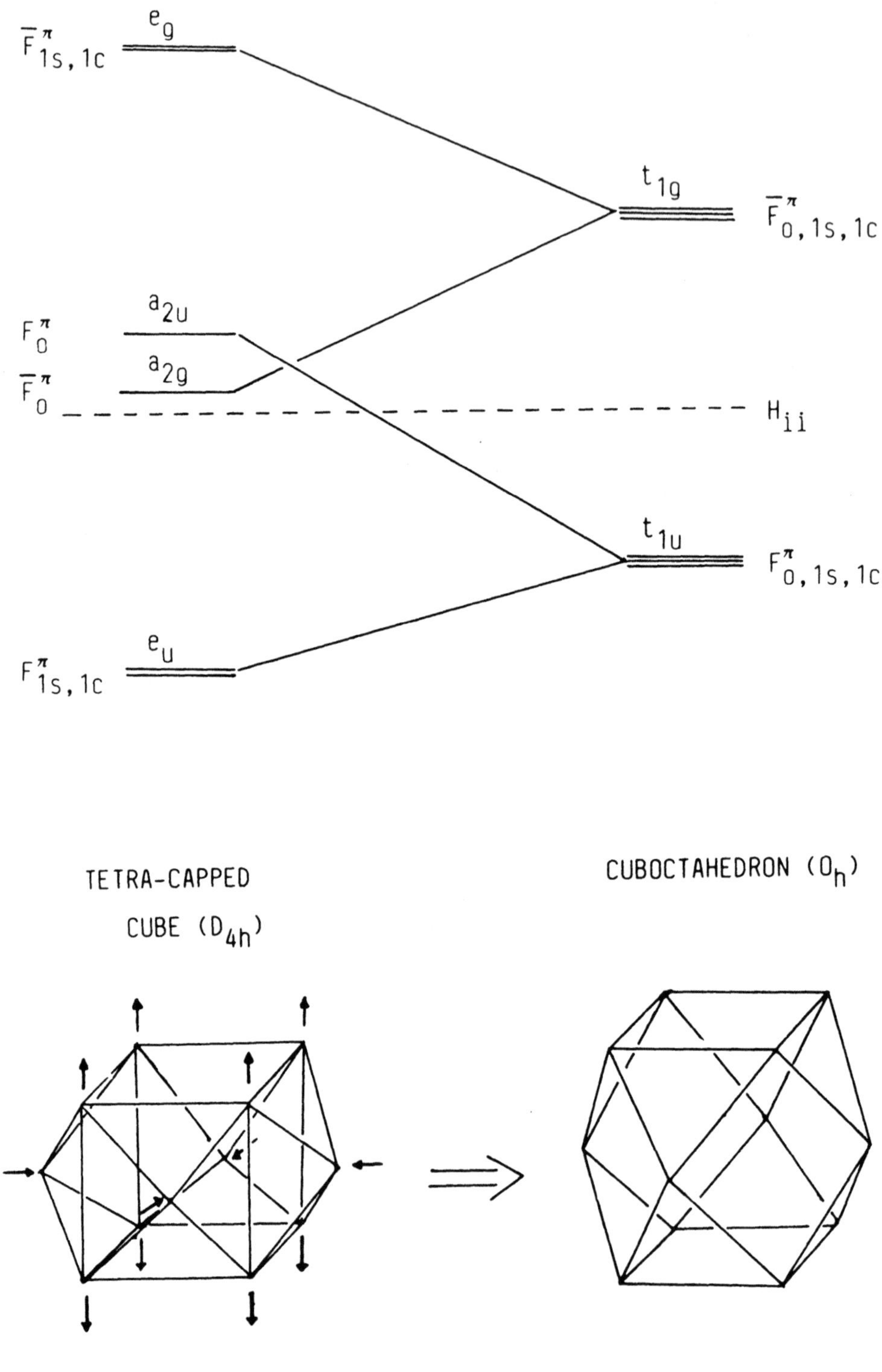

Fig. 23. Correlation of the frontier orbitals of the tetracapped cube and the cuboctahedron

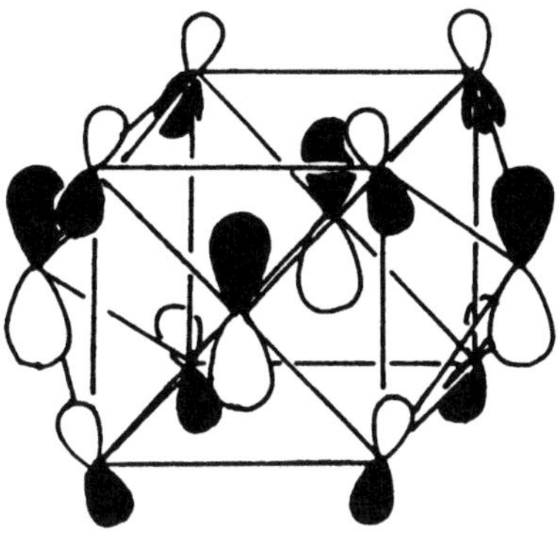

F_0^{π}

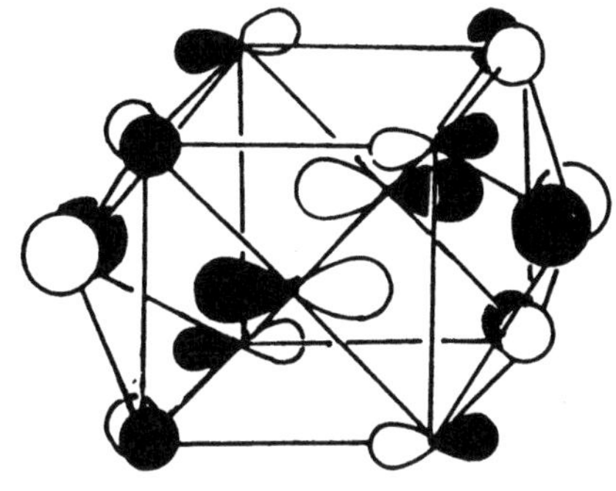

$\bar{F}_0^{\pi}$

Fig. 23 (continued)

Table 9. Bonding and non-bonding skeletal MO's of some 4-connected and related capped 3-connected polyhedral clusters

n	Cluster skeleton	Bonding skeletal Mo's	Non-bonding skeletal MO's[a]
6	Octahedron (O_h)	$S^{\sigma}(a_{1g})$, $P^{\sigma}(t_{1u})$, $D^{\pi}_{\pm1,2s}(t_{2g})$	
8	Square-antiprism (D_{4d})	$S^{\sigma}(a_1)$, $P^{\sigma}(b_2 + e_1)$, $D^{\pi}_0(a_1)$, $D^{\pi}_{\pm2}(e_2)$	$D^{\pi}_{\pm1}(e_3)$, $\bar{D}^{\pi}_{\pm1}(e_1)$
9	Tricapped trigonal prism (TTP) (D_{3h})	$S^{\sigma}(a_1')$, $P^{\sigma}(a_2'' + e')$, $D^{\pi}(a_1' + e' + e'')$	$F^{\pi}_0(a_2'')$, $\bar{F}^{\pi}_0(a_2')$[b]
9	Elongated TTP (D_{3h})	As for TTP but $F^{\pi}_0(a_2'')$ is bonding and $\bar{F}^{\pi}_0(a_2')$ is antibonding.	
10	Bicapped cube (BCC) D_{4h})	$S^{\sigma}(a_{1g})$, $P^{\sigma}(a_{2u} + e_u)$, $D^{\pi}_0(a_{1g})$, $D^{\pi}_{\pm1}(e_g)$, $D^{\pi}_{2s}(b_{2g})$, $F^{\pi}_{\pm1}(e_u)$	$D^{\pi}_{2c}(b_{1g})$, $\bar{D}^{\pi}_{2c}(b_{1u})$[b]
10	Distorted BCC[c] (D_{4h})	As for BCC but $D^{\pi}_{2c}(b_{1g})$ is bonding ($\bar{D}^{\pi}_{2c}(b_{1u})$ remains effectively non-bonding, however).	
12	Tetracapped cube (D_{4h})	$S^{\sigma}(a_{1g})$, $P^{\sigma}(a_{2u} + e_u)$, $D^{\pi}(a_{1g} + b_{1g} + b_{2g} + e_g)$, $F^{\pi}_{\pm1}(e_u)$, $F^{\pi}_{2c}(b_{1u})$	$F^{\pi}_0(a_{2u})$, $\bar{F}^{\pi}_0(a_{2g})$[b]
12	Cuboctahedron (O_h)	$S^{\sigma}(a_{1g})$, $P^{\sigma}(t_{1u})$, $D^{\pi}(e_g + t_{2g})$, $F^{\pi}(t_{1u} + a_{2u})$	
	(a) defined down C_4	$F^{\pi}(t_{1u})$, $F^{\pi}_0(a_{2u})$	
	(b) defined down C_3	$F^{\pi}(t_{1u})$, $F^{\pi}_{2c}(a_{2u})$	
12	Twinned-cuboctahedron (D_{3h})	$S^{\sigma}(a_1')$, $P^{\sigma}(a_2'' + e')$, $D^{\pi}(a_1' + e' + e'')$ $F^{\pi}_0(a_2'')$, $F^{\pi}_{\pm1}(e')$, $F^{\pi}_{3c}(a_1')$	

[a] Including weakly bonding and weakly antibonding skeletal MO's.

[b] In these non-spherical cases the (approximately) non-bonding $\bar{L}^{\pi}$ orbital lies at lower energy (is more strongly bonding) than its parity related L^{π} counterpart.

[c] The (D_{4h}) distorted bicapped cube has all of its vertices equidistant from the centre of the polyhedron (i.e. is "spherical") and possesses a regular hexagonal cross section in two perpendicular planes which contain the C_4 axis.

their TSH labels) for a number of 3-connected polyhedral geometries are listed in Table 10[152].

It is therefore, apparent that, despite the fact that the vertices of a 3-connected polyhedron may all lie on the surface of a single sphere, the arrangement of these vertices represents a poor coverage of the sphere's surface. Thus, these polyhedra do not emulate the sphere sufficiently well for their skeletal MO's to be good approximations to the particle on a sphere solutions. In particular, the low number of close atom-atom contacts (i.e. cluster edges) is insufficient to enable a distinction to be made, in all cases, of the bonding characteristics of the even (L^{π}) and odd parity ($\bar{L}^{\pi}$) tangential orbitals.

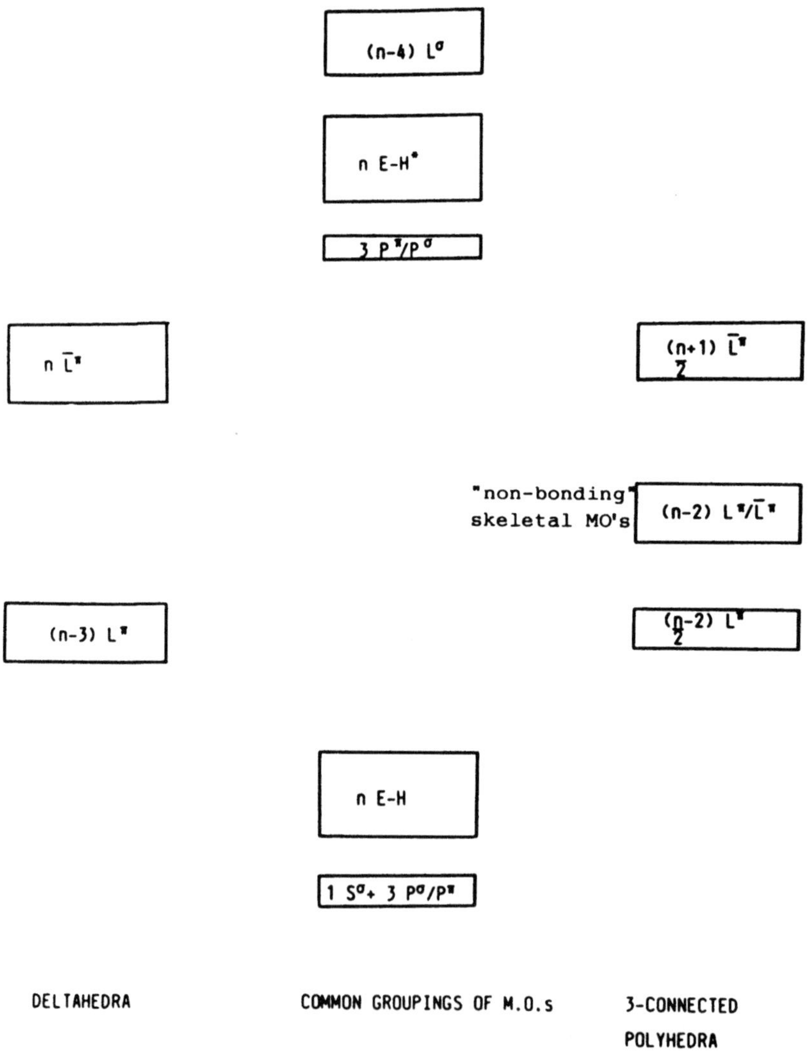

Fig. 24. Schematic comparison of the skeletal MO pattern of deltahedral and 3-connected $[E_nH_n]$ molecules (where E is a main group atom)

Table 10. Bonding and non-bonding skeletal MO's[a] of some 3-connected polyhedral clusters

n	Bonding skeletal MO's	Non-bonding skeletal MO's
4 (T_d)	$S^\sigma(a_1)$, $P^\sigma(t_2)$[b]	$D_0^\pi/\bar{D}_0^\pi(e)$
6 (D_{3h})	$S^\sigma(a_1')$, $P^\sigma(a_2'' + e')$, $D_0^\pi(a_1')$	$D_{\pm1}^\pi(e'')$, $\bar{D}_{\pm1}^\pi(e')$
8 (O_h)	$S^\sigma(a_{1g})$, $P^\sigma(t_{1u})$, $D_{0,2s}^\pi(e_g)$	$D_{\pm1,2c}^\pi(t_{2g})$, $\bar{D}_{\pm1,2c}^\pi(t_{2u})$
10 (D_{5h})	$S^\sigma(a_1')$, $P^\sigma(a_2'' + e_1')$, $D_0^\pi(a_1')$, $D_{\pm2}^\pi(e_2')$	$D_{\pm1}^\pi(e_1'')$, $F_{\pm2}^\pi(e_2'')$, $\bar{D}_{\pm1}^\pi(e_1')$, $\bar{F}_{\pm2}^\pi(e_2')$
12 (D_{6h})	$S^\sigma(a_{1g})$, $P^\sigma(a_{2u} + e_{1u})$, $D_0^\pi(a_{1g})$, $D_{\pm2}^\pi(e_{2g})$, $F_{3s}^\pi(b_{1u})$	$D_{\pm1}^\pi(e_{1g})$, $F_{\pm2}^\pi(e_{2u})$, $F_{3c}^\pi(b_{2u})$, $\bar{D}_{\pm1}^\pi(e_{1u})$, $\bar{F}_{\pm2}^\pi(e_{2g})$, $\bar{F}_{3c}^\pi(b_{2g})$
12 (T_d)	$S^\sigma(a_1)$, $P^\sigma(t_2)$, $D_{0\pm2}^\pi(t_2)$, $F_0^\pi(a_1)$	$D_{\pm1}^\pi(e)$, $F^\pi(t_1)$ $\bar{D}_{\pm1}^\pi(e)$, $\bar{F}^\pi(t_2)$
20 (I_h)	$S^\sigma(a_g)$, $P^\sigma(t_{1u})$, $D^\pi(h_g)$, $F^\pi(t_{2u})$	$F^\pi(g_u)$, $G^\pi(h_g)$, $\bar{F}^\pi(g_g)$, $\bar{G}^\pi(h_u)$

[a] The TSH labels refer to the case where the polar axis is the principal rotation axis, with the exception of the tetrahedron (n = 4), where the orbitals are labelled relative to the C_2 axis.

[b] The orbitals labelled P^σ (for convenience) are actually $P^{\sigma+\pi}$ (i.e. they possess a certain amount of tangential orbital character).

III. A Graph Theoretical Analogue of the Tensor Surface Harmonic Theory

Several years before the formulation of the Tensor Surface Harmonic Theory, King presented a graph theoretical treatment of bonding in borane clusters[56a)]. It is instructive to compare this with Stone's methodology.

King started by representing an n-atom *closo* cluster by the complete graph K_n, in which all the vertices are connected (for example, K_5 is shown in (**4**)). By applying algebraic graph theoretical methods[153)] and considering the topology of the bonding interactions of the atomic σ- und π-symmetry orbitals, King derived the eigenvalue spectra of K_n for a set of n radial (σ) atomic orbitals. He showed that the n σ orbitals gave rise to one positive eigenvalue (bonding cluster MO) and (n − 1) degenerate negative eigenvalues (antibonding cluster MO's). King also indicated that the 2n π orbitals gave rise to n degenerate bonding cluster MO's and n degenerate cluster antibonding MO's Wade's (n + 1) SEP rule was derived by assuming the occupation of all the bonding cluster MO's[56a)].

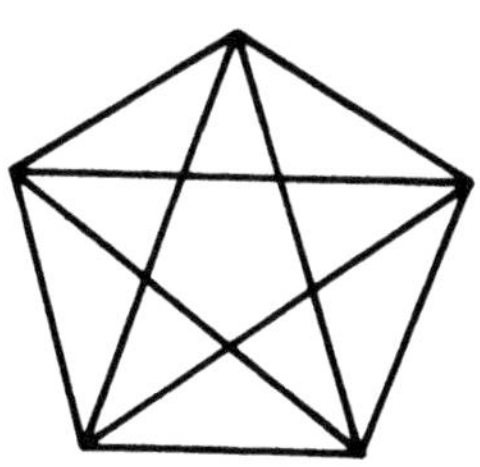

K_5

(**4**)

King's method (which he also used to derive electron counts for *nido* and *arachno* boranes and capped transition metal clusters[56a)] is essentially one step further from the real cluster geometry than Stone's TSH theory. Representing a cluster by the complete graph K_n is equivalent to considering the cluster as hyperspherical (as (n − 1) dimensions are required to make the interactions between n atoms equal). This explains why King's analysis leads to higher orbital degeneracies than those obtained from TSH theory.

IV. Point Group Symmetry Theorems Within the Tensor Surface Harmonic Framework

1. Introduction

Recently, Quinn and coworkers have developed a group theoretical formalism whereby symmetry adapted linear combinations of ligand orbitals are generated within the Tensor Surface Harmonic framework[154)] (all the ligands are taken to lie on the surface of a single sphere). This methodology has been shown to be equally applicable to cluster compounds, enabling the derivation of the L^{σ}, $L^{\pi}/\bar{L}^{\pi}$ and $L^{\delta}/\bar{L}^{\delta}$ orbitals by considering point group symmetries on the spherical shell[154)]. The approach makes use of Mollweide projections (used in cartography[155)] to display the nodal characteristics of the Spherical Harmonics in two dimensions. Fowler[156)] and ourselves[157)] have subsequently applied Quinn's group theoretical rules to the study of a variety of cluster molecules. A related approach has been developed by Ceulemans, who has used Induction Theory[158)] to consider the local site-symmetry of the cluster vertex atoms as well as the overall symmetry of the cluster[159)].

2. Group Theoretical Methodology

In Quinn's methodology, the reducible representations corresponding to linear combinations of s, p, d, f etc. atomic orbitals, located at the cluster vertices, may be denoted Γ_s, Γ_p, Γ_d, Γ_f etc. Quinn has shown that the reducible representation Γ_ℓ (ℓ = 0, (s), 1 (p), 2 (d) etc.) may be generated by multiplying the permutation representation of the cluster (Γ_ϱ) by the representations spanned by the $Y_{\ell,m}$ functions at the origin of the cluster (Γ_ℓ^0)[154)]:

$$\Gamma_\ell = \Gamma_\varrho \otimes \Gamma_\ell^0 \tag{28}$$

The (reducible) permutation representation (Γ_ϱ) is such that the character ($\chi_\varrho(R)$) for each operation (R) of the point group is equal to the number of cluster vertices which remain fixed under that operation[160)].

The reducible representation Γ_ℓ is clearly the sum of irreducible representations (Γ^i):

$$\Gamma_\ell = \sum_i a_i \Gamma^i \tag{29}$$

If the character of R in each irreducible representation is denoted by $\chi^i(R)$, then, by applying the group orthogonality principle[160)] the coefficients (a_i) (i.e. number of times that the ith irreducible representation appears in Γ_ℓ) are given by:

$$a_i = (1/h) \sum_R \chi^i(R)\chi_\varrho(R) \tag{30}$$

where h is the order of the group[160)].

Because s orbitals are nodeless, their linear combinations generate (in TSH terms) the cluster L_s^σ MO's[154]:

$$\Gamma_s = \Gamma_\varrho \otimes \Gamma_s^0 = \Gamma_\sigma \quad (\Gamma_\sigma = \Gamma(L^\sigma)) \tag{31}$$

where Γ_s^0 is the totally symmetry irreducible representation, Γ_g^0, with character +1 under all operations. Since multiplication by Γ_g^0 leaves any representation unchanged, $\Gamma_\sigma = \Gamma_\varrho$.

At the origin, the p atomic orbitals transform as $\Gamma_{x,y,z}$. Using the above equations, and the fact that $\Gamma_\varrho = \Gamma_\sigma$, enables the following relationship to be derived[154]:

$$\Gamma_p = \Gamma_\sigma \otimes \Gamma_p^0 = \Gamma_\sigma \otimes \Gamma_{x,y,z} \tag{32}$$

Unlike the s orbitals, the atomic p orbitals possess one angular node. At each cluster vertex, linear combinations of p_x, p_y, and p_z may be taken, which generate one radial (p^σ) and two tangential (p^π) orbitals. In terms of TSH theory these orbitals form the basis of the cluster L_p^σ and $L^\pi{}_p/\bar{L}_p^\pi$ MO's. In group theoretical terms, Γ_p can be divided into radial and tangential parts as follows:

$$\Gamma_p = \Gamma_\sigma + \Gamma_{\pi+\bar{\pi}} \tag{33}$$

($\Gamma_{\pi+\bar{\pi}} = \Gamma_\pi + \Gamma_{\bar{\pi}}$; $\Gamma_\pi = \Gamma(L^\pi)$, $\Gamma_{\bar{\pi}} = \Gamma(\bar{L}^\pi)$; $\Gamma_{\bar{\pi}} = \Gamma_\pi \otimes \Gamma_u^0$.) Γ_u^0 is the antisymmetric pseudo scalar representation, with character +1 for proper rotations and −1 for improper rotations, including inversion, S_n axes and reflections. The symmetries of the Spherical Harmonics (Y_{LM}) and Vector Surface Harmonics ($V_{LM}/\bar{V}_{LM}$), with $L \leqslant 4$ are correlated in Table 11 for the spherical group (R_{3h}) and some finite point groups.

Table 11. Symmetries of the even and odd parity Spherical Harmonics (with L < 4) in some finite point groups[a]

L	R_{3h}	I_h	O_h	T_d	D_3
(a) Even Parity					
0	S_g^σ (Γ_g^0)	A_g	A_{1g}	A_1	A_1
1	P_u^σ, P_u^π	T_{1u}	T_{1u}	T_2	$A_2 + E$
2	D_g^σ, D_g^π	H_g	$E_g + T_{2g}$	$E + T_2$	$A_1 + 2E$
3	F_u^σ, F_u^π	$T_{2u} + G_u$	$A_{2u} + T_{1u} + T_{2u}$	$A_1 + T_1 + T_2$	$A_1 + 2A_2 + 2E$
4	G_g^σ, G_g^π	$G_g + H_g$	$A_{1g} + E_g + T_{1g} + T_{2g}$	$A_1 + E + T_1 + T_2$	$2A_1 + A_2 + 3E$
(b) Odd Parity					
0	Γ_u^0	A_u	A_{1u}	A_2	A_1
1	$\bar{P}_g^\pi$	T_{1g}	T_{1g}	T_1	$A_2 + E$
2	$\bar{D}_u^\pi$	H_u	$E_u + T_{2u}$	$E + T_1$	$A_1 + 2E$
3	$\bar{F}_g^\pi$	$T_{2g} + G_g$	$A_{2g} + T_{1g} + T_{2g}$	$A_2 + T_1 + T_2$	$A_1 + 2A_2 + 2E$
4	$\bar{G}_u^\pi$	$G_u + H_u$	$A_{1u} + E_u + T_{1u} + T_{2u}$	$A_2 + E + T_1 + T_2$	$2A_1 + A_2 + 3E$

[a] The representations spanned by the even parity Spherical Harmonics (Y_L, $L \geqslant 2$), in a given point group, may be obtained from the following recursive formula:

$$\Gamma_{L+1} = \{\Gamma_L \otimes \Gamma_{x,y,z}\} - \{\Gamma_L \otimes \Gamma_u^0\} - \Gamma_{L-1}$$

(e.g. $\Gamma_D = \{\Gamma_P \otimes \Gamma_{x,y,z}\} - \{\Gamma_P \otimes \Gamma_u^0\} - \Gamma_S$). The representations spanned by the odd parity functions are then obtained by multiplying the even parity representations by Γ_u^0:

$$\bar{\Gamma}_{L+1} = \Gamma_{L+1} \otimes \Gamma_u^0$$

Similarly, for the d based cluster MO's the following relationships may be written:

$$\Gamma_d = \Gamma_\sigma \otimes \Gamma_d^0(z^2, x^2 - y^2, xy, xz, yz)$$
$$= \Gamma_\sigma + \Gamma_{\pi+\bar{\pi}} + \Gamma_{\delta+\bar{\delta}} \tag{34}$$
$$(\Gamma_{\delta+\bar{\delta}} = \Gamma_\delta + \Gamma_{\bar{\delta}}, \Gamma_{\bar{\delta}} = \Gamma_\delta \otimes \Gamma_u^0)$$

For f based cluster MO's:

$$\Gamma_f = \Gamma_\sigma + \Gamma_{\pi+\bar{\pi}} + \Gamma_{\delta+\bar{\delta}} + \Gamma_{\phi+\bar{\phi}} \tag{35}$$

Quinn has used these relationships to obtain equations for deriving the symmetries of cluster π- and δ-type MO's from those of the L^σ orbitals[154]:

$$\Gamma_{\pi+\bar{\pi}} = \{\Gamma_\sigma \otimes \Gamma_{x,y,z}\} - \Gamma_\sigma$$
$$\Gamma_{\delta+\bar{\delta}} = \{\Gamma_{\pi+\pi} \otimes \Gamma_{x,y,z}\} - \{\Gamma_\sigma \otimes \Gamma_u^0\} - \Gamma_\sigma \tag{36}$$

In general, if m indicates the number of angular nodes of an atomic orbital (at a cluster vertex) relative to the radial vector, and Γ_m is the reducible representative spanned by linear combinations of these orbitals ($\Gamma_0 = \Gamma_\sigma$, $\Gamma_1 = \Gamma_{\pi+\bar{\pi}}$, $\Gamma_2 = \Gamma_{\delta+\bar{\delta}}$ etc.), then the following relationship can be derived[154]:

$$\Gamma_{m+1} = \{\Gamma_m \otimes \Gamma_{x,y,z}\} - \Gamma_m - \Gamma_{m-1} \tag{37}$$

This equation is valid for $m \geqslant 2$ (e.g. $\Gamma_{\phi+\bar{\phi}} = \{\Gamma_{\delta+\bar{\delta}} \otimes \Gamma_{x,y,z}\} - \Gamma_{\delta+\bar{\delta}} - \Gamma_{\pi+\bar{\pi}}$).

Quinn has derived simplified forms of Eqs. (36) and (37) for axial point groups, where the p^θ and p^ϕ components of the tangential p^π orbitals are seperable. The tangential orbitals are merely formed by taking the L^σ MO's with a basis set of p^θ and p^ϕ orbitals, rather than radial orbitals. The p^θ orbitals have the same local symmetry as Γ_z, while the p^ϕ orbitals have the local symmetry of Γ_{Rz}[154]:

$$\Gamma_{\pi+\bar{\pi}} = \Gamma^\theta_{\pi+\bar{\pi}} + \Gamma^\phi_{\pi+\bar{\pi}}$$
$$= \{\Gamma_\sigma \otimes \Gamma_z\} + \{\Gamma_\sigma \otimes \Gamma_{Rz}\} \tag{38}$$

Similarly, Quinn has shown that, in axial point groups, the L^δ and $\bar{L}^\delta$ orbitals can be split into their $d^{\theta\theta}$ and $d^{\theta\phi}$ components. The former have the same local symmetry as Γ_{z^2} (which transforms as Γ_g^0 in axial point groups), while the latter transform as Γ_u^0[154]:

$$\Gamma_{\delta+\bar{\delta}} = \{\Gamma_\sigma \otimes \Gamma_{z^2}\} + \{\Gamma_\sigma \otimes \Gamma_u^0\}$$
$$= \Gamma_\sigma + \{\Gamma_\sigma \otimes \Gamma_u^0\} \tag{39}$$

Equations (38) and (39) are not, however, valid for cluster polyhedra possessing vertices (i.e. atoms) which lie on the principal rotation axis, where p_z and d_{z^2} have purely radial character. In these cases, provided there are no atoms lying in the equatorial plane, we have noted that the following equations may be employed[157b)]:

$$\Gamma_{\pi+\bar{\pi}} = \Gamma_\sigma \otimes \Gamma_{x,y} \tag{40}$$

$$\Gamma_{\delta+\bar{\delta}} = \Gamma_\sigma \otimes \Gamma_{xy,x^2-y^2} \tag{41}$$

These follow because, at all cluster vertex positions (except those in the equatorial plane), linear combinations of p_x and p_y (or d_{xy} and $d_{x^2-y^2}$) may be taken to generate two tangential p^π (or d^δ) orbitals.

3. *Application of the Symmetry Theorems to the Skeletal Molecular Orbitals of Octahedral and Tetrahedral Clusters*

To illustrate some of the points outlined above the following is a derivation of the symmetries of the L^σ, L^π and $\bar{L}^\pi$ orbitals of a general 6-vertex octahedral cluster and a 4-vertex tetrahedral cluster.

For the octahedron, by standard Group Theoretical methods[160)]:

$$\Gamma_\sigma = \Gamma_\varrho = a_{1g}(S^\sigma) + t_{1u}(P^\sigma) + e_g(D^\sigma)$$

The L labels have been obtained from a correlation table such as Table 11. In the O_h point group $\Gamma_{x,y,z} = t_{1u}$ so, by applying Eq. (36) and consulting a table of direct products:

$$\Gamma_{\pi+\bar{\pi}} = \{\Gamma_\sigma \otimes t_{1u}\} - \Gamma_\sigma = t_{1u} + t_{2g} + t_{1g} + t_{2u}$$

the L values of these tangential orbitals and their parity (i.e. even or odd) may also be obtained from Table 11. These orbitals form two sets. L^π and $\bar{L}^\pi$, which are strictly bonding and antibonding respectively. They are related by the Parity Inversion operation ($\hat{\mathcal{P}}$), which transforms as Γ_u^0 (i.e. a_{1u} in O_h):

$$P^\pi(t_{1u}) \xleftrightarrow{(\hat{\mathcal{P}})\,a_{1u}} \bar{P}^\pi(t_{1g})$$

$$D^\pi(t_{2g}) \longleftrightarrow \bar{D}^\pi(t_{2u})$$

The mixing of L^σ and L^π orbitals with the same L values and symmetries results in a total of 7 (n + 1) skeletal bonding cluster MO's ($S^\sigma(a_{1g}) + P^{\sigma+\pi}(t_{1u}) + D^\pi(t_{2g})$).

Considering the tetrahedron:

$$\Gamma_\sigma = \Gamma_\varrho = a_1(S^\sigma) + t_2(P^\sigma)$$

In the T_d point group, $\Gamma_{x,y,z} = t_2$, so:

$$\Gamma_{\pi+\bar{\pi}} = \{\Gamma_\sigma \otimes t_2\} - \Gamma_\sigma = t_1 + t_2 + e$$

Fowler has shown that in this point group $\Gamma_u^0 = a_2$ so the t_1 and t_2 orbitals can be seen to be parity-related (in group theoretical terms the representations t_1 and t_2 are said to be ***"conjugate"***[156a)] since $t_1 \otimes a_2 = t_2$ and $t_2 \otimes a_2 = t_1$. The e representation, however, is ***"self-conjugated"*** since $e \otimes a_2 = e$. Since there is only one e representation in $\Gamma_{\pi+\pi}$, Fowler noted that this must correspond to a degenerate pair of $L^\pi/\bar{L}^\pi$ orbitals which, following Stone's argument[145c)] must consequently, be non-bonding. By consulting Table 11, the following L labels may be assigned to the tangential orbitals:

$$P^\pi(t_2) \xleftrightarrow{\hat{\mathscr{P}}} \bar{P}^\pi(t_1)$$

$$\hat{\mathscr{P}} \circlearrowleft \quad D^\pi/\bar{D}^\pi \quad (e)$$

The possible occupation of the frontier orbital $D^\pi/\bar{D}^\pi$ pair results in possible SEP counts of n (4) or (n + 2) (6), but not (n + 1) for the tetrahedral clusters. Generally these clusters are characterised by 6 SEP's (as in, for example, $[C_4Bu^t_4]$[161)] in keeping with their designation as electron-precise 3-connected clusters and *nido*-trigonal bipyramids, but examples are also known which possess 4 SEP's (e.g. $[B_4Cl_4]$[162)]).

4. *Further Applications of the Symmetry Theorems*

(a) Deviations from the (n + 1) Skeletal Electron Pair Rule for Deltahedral Clusters

Fowler has utilised the ideas presented above to rationalise, from a group theoretical standpoint, departures from the (n + 1) SEP rule which have been predicted (from Lipscomb's LCAO-MO calculations[163)]) for certain hypothetical high nuclearity deltahedral boranes[156a)]. He has concluded that symmetry-induced departures occur, as in the case of the tetrahedron mentioned above, when $\Gamma_{\pi+\bar{\pi}}$ possess an odd number of self-conjugate e representations. This results in a number of bonding (L^π) e pairs, an equal number of antibonding ($\bar{L}^\pi$) e pairs and one non-bonding ($L^\pi/\bar{L}^\pi$) pair. The work of Fowler[156a)], as well as our own[157a, 146)], has shown that, for deltahedra, the occurence of an odd number of self-conjugate representations occurs as a consequence of there being

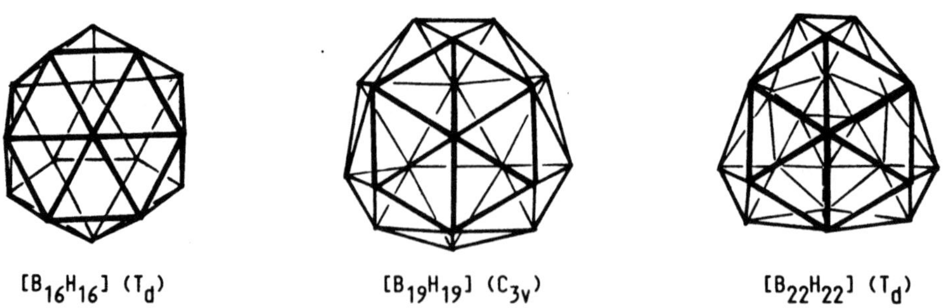

$[B_{16}H_{16}]$ (T_d) $[B_{19}H_{19}]$ (C_{3v}) $[B_{22}H_{22}]$ (T_d)

Fig. 25. Postulated high nuclearity borane clusters, with one cluster vertex on each C_3 rotation axis, which are predicted to disobey the (n + 1) SEP rule one atom on each C_3 rotation axis of a polyhedron belonging to one of the point groups C_3, C_{3v}, T or T_d. Examples of high nuclearity boranes with such geometries are shown in Fig. 25. We have also shown[157a, 164] that the same ideas may be applied to explain the hyperelectron deficiency of the n-vertex-n-SEP ***"hyper-closo"*** metalloboranes such as $[H(PPh_3)(Ph_2PC_6H_4)IrB_9H_8]$ **(5)**[165].

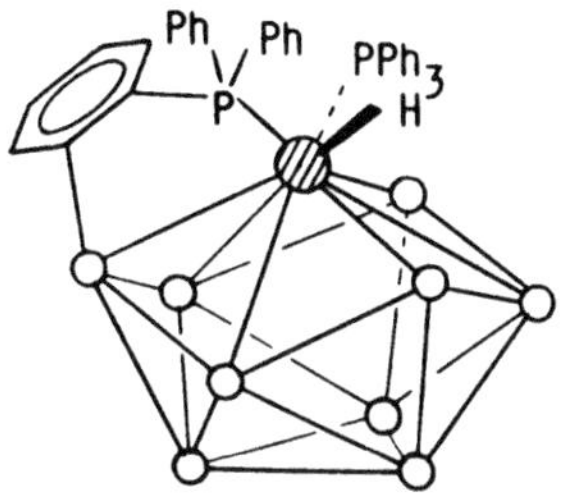

$[H(PPh_3)(Ph_2PC_6H_4)IrB_9H_8]$

(5)

We have developed a classification scheme for deltahedra which enables the prediction of the nature of any departures from the (n + 1) rule in terms of topological features of the cluster polyhedron[157a, 163]. Deltahedra may be classified according to whether they are "centrosymmetric spherical" (i.e. pseudo-spherical, belonging to O_h, or I_h point groups) in which case there is always a symmetrical and bonding distinction between the L^{π} and the $\bar{L}^{\pi}$ orbitals and the (n + 1) rule is rigorously obeyed, or "non-spherical". These non-spherical deltahedra may be termed "non-polar", "polar" or "bipolar", depending upon whether they possess 0, 1 or 2 vertices respectively on their principal C_n rotation axis (axes). The tetrahedron, the high nuclearity boranes depicted in Fig. 25 and the *hyper-closo* metalloboranes such as **(5)** are all examples of "C_3-polar deltahedra", having one vertex on each C_3 (principal) axis and, consequently, possessing a non-bonding degenerate $L^{\pi}/\bar{L}^{\pi}$ pair and an SEP count of n or (n + 2)[164].

The bipolar (e.g. bipyramids and bicapped antiprisms) and non-polar (e.g. D_{2d}-dodecahedron and tricapped trigonal prism) deltahedral custers are best analysed in terms of the interactions between the two sets (polar and non-polar or equatorial and non equatorial) of symmetry-equivalent atoms which make up the cluster[157a]. In this way it has been shown that, although by symmetry there are no degenerate $L^{\pi}/\bar{L}^{\pi}$ pairs and therefore no symmetry-induced departures from the (n + 1) rule, the frontier orbitals of bipolar deltahedral clusters consist of two parity matched L^{π} and $\bar{L}^{\pi}$ e pairs, giving rise to possible SEP counts of (n − 1), (n + 1) or (n + 3). In the case of the non-polar deltahedra the frontier orbitals are non-degenerate L^{π} and $\bar{L}^{\pi}$ orbitals, giving rise to possible SEP counts of n, (n + 1) or (n + 2).

(b) Bispherical and Multispherical Clusters

In an extension of this work, we have considered the bonding in bispherical and multispherical clusters where the skeletal atoms lie on the surfaces of two or more concentric spheres[157b]. This type of model may be used to analyse face-capped clusters. In bispherical cases, it was found that the presence of a ring (either planar or puckered) around the equator of a cluster (thereby defining a toroidal topology) leads to the stabilisation of one (initially antibonding) skeletal MO of the central (inner sphere) polyhedron. If, however, the capping atoms define a 3-dimensional, pseudospherical coverage (as in omnicapped polyhedra) then three additional skeletal MO's are stabilised.

E. Further Aspects of Theoretical Cluster Chemistry

I. The Application of Sophisticated Calculational Techniques to Problems of Cluster Bonding

The preceding sections have demonstrated that empirical and semi-empirical calculational techniques, such as the Extended Hückel method[166)], have made considerable contributions to our understanding of the bonding in main group and transition metal cluster molecules. As the quantity of spectroscopic data on clusters has grown, however, more sophisticated calculational techniques have been employed in an attempt to reproduce experimental observables and in so doing to gain a more detailed understanding of the electronic structures of these molecules. The electronic structures of bare metal and ligated clusters have been investigated using the theoretical methodologies of solid state physics (e.g. chemical pseudopotential approaches[167a)] and Xα methods[127, 167b)]) in addition to the more familiar (to the chemist) LCAO-MO methods of varying degress of sophistication (e.g. *ab initio*[168)], Fenske-Hall[169)] and CNDO[170)]. A comprehensive review of spectroscopic and theoretical studies of transition metal cluster compounds has been published by Manning and Trogler[171)].

II. From Discrete Cluster Molecules to Metal-Metal Bonds in the Solid State

In recent years, a considerable amount ot synthetic and theoretical attention has been focussed on extended metal-metal bonded systems in the solid state, inspired by the interesting physical properties which some of these compounds exhibit[15)] and the wealth of structural data available[13)]. Theoretical studies may be divided into those using essentially free electron models (notably by Andersen[172)]) and the tight binding calculations favoured by Hoffman, Burdett and co-workers[173)].

III. The Problem of the Arrangement of Ligands Around the Cluster Core

Although most theoretical treatments of transition metal cluster compounds have concentrated on metal skeletal geometry-electron count correlations, an additional (often more difficult) problem concerns the alternative arrangements adopted by the ligands (especially carbonyl ligands) and ligand fluctionality. Dahl noted that the carbonyl ligands in the triangular cluster $[Fe_3(CO)_{12}]$ adopt an icosahedral arrangement and employed a hard sphere packing model to rationalise this fact[174]. Johnson and Benfield later developed a more elegant methodology for determining the most stable arrangements of ligands in clusters. In their treatment, the cluster and ligand polyhedra are matched in such a way that interligand repulsions are minimised and metal-ligand bonding is maximised[175]. Gillespie likewise analysed the problem in terms of the packing of carbonyl ligands around a cluster core[51] and similar arguments have subsequently been propounded by Kepert[176] and Mingos (who developed the concept of a Cluster Cone Angle in order to analyse steric interactions in gold phosphine clusters[177]. Lauher[178] and Kepert[179] have recently developed Surface Force Field Models for the Molecular Mechanics[180] simulation of ligand arrangements in metal carbonyl clusters.

There has been considerable debate as to the relative importance of steric and electronic factors in determining ligand arrangements in clusters[181]. Evans has suggested that electronic features of the constituent ML_n fragments may be used to explain, for example, the different structures adopted by the ligands in the tetrahedral clusters $[Co_4(CO)_{12}]$ (**6**) and $[Ir_4(CO)_{12}]$ (**7**)[181, 182]. Minot has used a similar metal fragment approach to analyse the problem of terminal versus bridging coordination of carbonyl ligands[183].

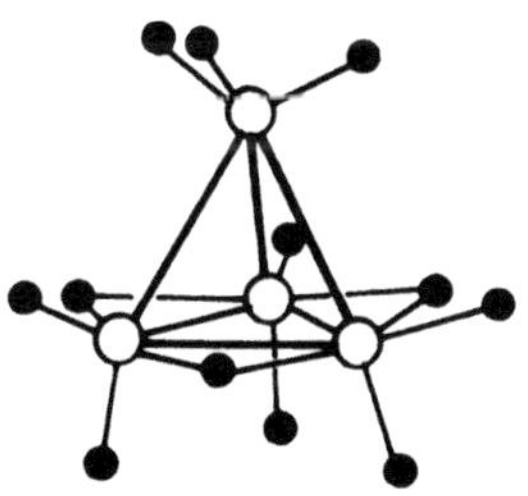

$[Co_4(CO)_{12}]$

(6)

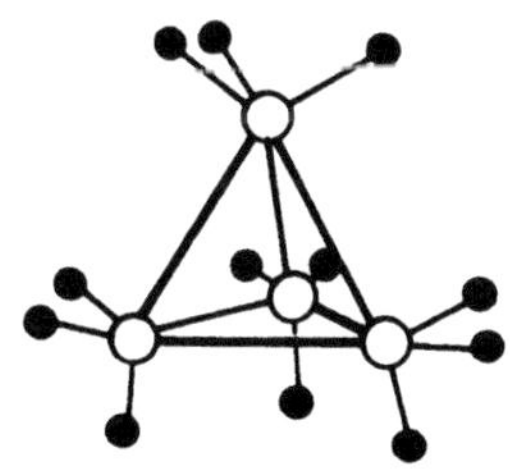

$[Ir_4(CO)_{12}]$

(7)

F. Summary

This article has outlined the crucial stages in the development of cluster bonding theories. On the whole it has been the qualitative and semi-empirical approaches which have led to the elucidation of patterns and the establishment of bonding models. This is partly because such methods are fast for cluster compounds. The success of topological and group theoretical approaches,which concentrate on the cluster polyhedra without being too concerned with the actual nature of the cluster atoms, has led to the realisation that most of the cluster structure-electron count correlations are symmetry and topology dependent. This explains the success of symmetry based semi-empirical approaches (such as the Extended Hückel method) in rationalising and predicting cluster geometries and valence electron counts[166].

It is difficult to predict with the ways in which cluster bonding theory will develop in the future, since the theoretical chemist is dependent on synthetic chemists, spectroscopists and crystallographers for raw data. As novel types of clusters are characterised, our ideas concerning the electronic structures of cluster molecules, have to be modified. The increasing availability of larger, faster (and cheaper!) computers should eventually enable the application of *ab-initio* calculational techniques to even the largest of clusters. It is likely that significant advances will be made in the area of rationalising and predicting physical properties of metal clusters (particularly in the solid state) in an effort to aid the synthetic chemist in his endeavour to develop new materials with tailor-made magnetic, conducting and catalytic properties.

Acknowledgement. The S.E.R.C. is thanked for financial support.

G. References

1. (a) Muetterties, E. L.: Chem. Eng. News, 30th August, 1982, 28; (b) Transition Metal Clusters (ed. Johnson, B. F. G.) Wiley, New York 1981 and references therein
2. (a) Lipscomb, W. N.: Boron Hydrides, Benjamin, New York 1963; (b) Boron Hydride Chemistry (ed. Muetterties, E. L.), Academic, New York 1975 and references therein. (c) The first boranes were characterised by Stock and coworkers: Stock A.: Hydrides of Boron and Silicon, Cornell University Press, Ithaca, New York 1933 and references therein. (d) For reviews of the related metallaboranes and metallacarboranes see: Metal Interactions with Boron Clusters (ed. Grimes, R. N.), Plenum, New York 1982; Grimes R. N.: Acc. Chem. Res. *16,* 22 (1983)
3. See, for example: Weilmuenster, E. A.: Ind. Eng. Chem. *49,* 1340 (1957); Martin, D. R.: J. Chem. Educ. *36,* 208 (1959)
4. Corbett, J. D.: Prog. Inorg. Chem. *21,* 129 (1976); Corbett, J. D.: Chem. Rev. *85,* 383 (1985); von Schnering, H. G.: Angew. Chem. Int. Edn. Engl. *20,* 33 (1981) and references therein. For theoretical treatments of naked metal clusters see: Belin, C. H. E., Corbett, J. D., Cisar, A.: J. Amer. Chem. Soc. *99,* 7163 (1977); Lohr, L. L.: Inorg. Chem. *20,* 4229 (1981); Burns, R. C., Gillespie, R. J., Barnes, J. A., McGlinchey, M. J.: Inorg. Chem. *21,* 799 (1982)
5. Paquette, L. A.: Acc. Chem. Res. *4,* 280 (1971); Clark, T., McKervey, M. A.: Comp. Organic Chem. *1,* 35 (1979); Ternansky, R. J., Balogh, D. W., Paquette, L. A.: J. Amer. Chem. Soc. *104,* 4503 (1982) and references therein
6. For recent reviews of catalytic processes which may involve transition metal cluster compounds see: Whyman, R.: in Ref. 1b, Ch. 8 p. 545; Whyman, R.: Phil. Trans. R. Soc. Lond. *A308,* 131 (1982) and references therein
7. Johnson, B. F. G., Lewis, J.: Adv. Inorg. Chem. Radiochem. *24,* 225 (1981); Johnson, B. F. G., Lewis, J.: Phil. Trans. R. Soc. Lond. *A308,* 5 (1982)
8. Muetterties, E. L., Rhodin, T. N., Band, E., Brucker, C. F., Pretzer, W. R.: Chem. Rev. *29,* 91 (1979); Muetterties, E. L., Wexler, R. M.: Survey Prog. Chem. *10,* 61 (1983)
9. Schmid, G.: Struct. Bond. *62,* 51 (1985) and references therein
10. Mingos, D. M. P., Wardle, R. W. M.: Transition Met. Chem. *10,* 441 (1985) and references therein. See also: Longoni, G., Chini, P.: J. Amer. Chem. Soc. *98,* 7225 (1976); Gregson, D., Howard, J. A. K., Murray, M., Spencer, J. L.: J. Chem. Soc. Chem. Commun. 716 (1981); Frost, P. W., Howard, J. A. K., Spencer, J. L., Turner, D. G.: J. Chem. Soc., Chem. Commun. 1104 (1981)
11. Hall, K. P., Mingos, D. M. P.: Prog. Inorg. Chem. *32,* 237 (1984) and references therein
12. Schäfer, H., von Schnering, H. G.: Angew. Chem. *76,* 833 (1964); Baird, M. C.: Prog. Inorg. Chem. *9,* 1 (1968); McCarley, R. E.: Phil. Trans. R. Soc. Lond. *A308,* 141 (1982) and references therein
13. Simon, A.: Angew. Chem. Int. Edn. Engl. *14,* 239 (1981); Corbett, J. D.: Acc. Chem. Res. *14,* 239 (1981)
14. Chevrel, R., Sergent, M., Prigent, J.: J. Solid State Chem. *3,* 515 (1971); Chevel, R., Gougeon, P., Potel, M., Sergent, M.: J. Solid State Chem. *57,* 25 (1985)
15. Matthias, B. T., Marezio, M., Corenzwit, E., Cooper, A. S., Barz, H. E.: Science *175,* 1465 (1972); Fischer, O.: Appl. Phys. *16,* 1 (1978); Topics in Current Physics: Superconductivity in Ternary Compounds (eds. Fischer, O., Maple, M. B.) Springer-Verlag, Berlin 1982, 32

16. Garner, C. D.: in Ref. 1b, Ch. 4 p. 235 and references therein
17. See, for example: Heaton, B. T.: Phil. Trans. R. Soc. Lond. *A308,* 95 (1982); Green, J. C., Seddon, E. A., Mingos, D. M. P.: J. Chem. Soc., Chem. Commun. 94 (1979)
18. Raithby, P. R.: in Ref. 1b, Ch. 2 p. 5
19. Heitler, H., London, F.: Z. Phys. *44,* 455 (1927)
20. Pauling, L.: The Nature of the Chemical Bond (3rd. Edn.), Cornell University Press, Ithaca, New York 1960
21. Mulliken, R. S.: Phys. Rev. *41,* 49 (1932)
22. Pauling, L., Wilson, E. B.: Introduction to Quantum Mechanics, McGraw-Hill, New York 1935 and references therein
23. Wade, K.: Electron Deficient Compounds, Nelson, London 1971
24. Longuet-Higgins, H. C.: J. Chim. Phys. *46,* 275 (1949)
25. Longuet-Higgins, H. C.: J. Roy. Inst. Chem. *77,* 197 (1953)
26. Duffey, G. H.: J. Chem. Phys. *19,* 963 (1951); *20,* 195 (1952); *21,* 761 (1953)
27. Eberhardt, W. H., Crawford Jr., B., Lipscomb, W. N.: J. Chem. Phys. *22,* 989 (1954)
28. Torkington, P.: J. Chem. Phys. *19,* 528 (1951)
29. Lennard-Jones, J. E.: Proc. Roy. Soc. A *198,* 1, 14 (1949)
30. Pople, J. A.: Quart. Rev. Chem. Soc. *11,* 273 (1957)
31. Switkes, E., Stevens, R. M., Lipscomb, W. N., Newton, M. D.: J. Chem. Phys. *51,* 2085 (1969); Switkes, E., Lipscomb, W. N., Newton, M. D.: J. Amer. Chem. Soc. *92,* 3847 (1970); Epstein, I. R., Tossell, J. A., Switkes, J. E., Stevens, R. M., Lipscomb, W. N.: Inorg. Chem. *10,* 171 (1971); Marynick, D. S., Lipscomb, W. N.: J. Amer. Chem. Soc. *94,* 1748 (1972); Laws, E. A., Stevens, R. M., Lipscomb, W. N.: J. Amer. Chem. Soc. *94,* 4467 (1972)
32. Kettle, S. F. A., Tomlinson, V.: J. Chem. Soc. (A) 2002 (1969)
33. Kettle, S. F. A., Tomlinson, V.: J. Chem. Soc. (A) 2007 (1969)
34. (a) Cotton, F. A.: Quart. Rev. Chem. Soc. *20,* 389 (1966) and references therein. (b) Lewis, J.: Pure Appl. Chem. *10,* 11 (1965); Chini, P.: Inorg. Chim. Acta. Rev. *2,* 31 (1968); Abel, E. W., Stone, F. G. A.: Quart. Rev. Chem. Soc. *23,* 325 (1969)
35. Vaughan, P. A.: Proc. Natl. Acad. Sci. *36,* 461 (1950)
36. Sheldon, J. C.: J. Chem. Soc. 1007, 3106 (1960)
37. Crossmann, L. D., Olsen, D. P., Duffey, G. H.: J. Chem. Phys. *38,* 73 (1963)
38. Cotton, F. A., Haas, T. E.: Inorg. Chem. *3,* 10 (1964)
39. Kettle, S. F. A.: Theor. Chim. Acta. *3,* 211 (1965)
40. Simon, A., von Schnering, H. G., Wohrle, H.: Z. Anorg. Allg. Chem. *339,* 155 (1965); Simon, A., von Schnering, H. G., Schäfer, H.: Z. Anorg. Allg. Chem. *361,* 235 (1968)
41. Kettle, S. F. A.: J. Chem. Soc. (A) 1013 (1966)
42. Maxwell, L. R., Hendricks, S. B., Mosley, V. M.: J. Chem. Phys. *3,* 699 (1935)
43. Kettle, S. F. A.: Theor. Chim. Acta. *4,* 150 (1966)
44. Sidgwick, N. V.: Electronic Theory and Valence, Clarendon Press, Oxford 1927; Craig, D. P., Doggett, G.: J. Chem. Soc. 4189 (1963); Tolman, C. A.: Chem. Soc. Rev. *1,* 337 (1972)
45. Sidgwick, N. V., Powell, H. E.: Proc. Roy. Soc. A *176,* 153 (1940)
46. Corey, E. R., Dahl, L. F., Beck, W.: J. Amer. Chem. Soc. *85,* 1202 (1963)
47. Kettle, S. F. A.: J. Chem. Soc. (A) 314 (1967)
48. Schulman, J. M., Fischer, C. R., Solomon, P., Venanzi, T. J.: J. Amer. Chem. Soc. *100,* 2949 (1978)
49. Heilbronner, E., Jones, T. B., Krebs, A., Malsch, D.-D., Maier, G., Pocklington, J., Schmelzer, A.: J. Amer. Chem. Soc. *102,* 564 (1980)
50. Mingos, D. M. P.: Nature Phys. Sci. *236,* 99 (1972)
51. Gillespie, R. J.: Chem. Soc. Rev. *8,* 315 (1979) and references therein
52. Gillespie, R. J., Nyholm, R. S.: Quart. Rev. Chem. Soc. *11,* 339 (1957); Gillespie, R. J.: Molecular Geometry, Van Nostrand-Reinhold, Princeton, New Jersey 1972
53. Burns, R. C., Gillespie, R. J., Luk, W.-C., Slim, D. R.: J. Chem. Soc., Chem. Commun. 791 (1976); Inorg. Chem. *18,* 3086 (1979)
54. Wei, C.-H., Dahl, L. F.: J. Amer. Chem. Soc. *88,* 1821 (1966); Strouse, C. E., Dahl, L. F.: Discuss. Faraday Soc. *47,* 93 (1969); Stevenson, D. L., Wei, C.-H., Dahl, L. F.: J. Amer. Chem. Soc. *93,* 6027 (1971); Strouse, C. E., Dahl, L. F.: J. Amer. Chem. Soc. *93,* 6032 (1971)
55. McGlinchey, M. J., Tal, Y.: Chemical Applications of Topology and Graph Theory, Studies in Phys. and Theor. Chem. *28,* 124 (1983)

56. (a) King, R. B., Rouvray, D. H.: J. Amer. Chem. Soc. *99,* 7834 (1977). (b) King, R. B.: Inorg. Chim. Acta. *49,* 237 (1981); *57,* 79 (1982); *116,* 99, 109, 119, 125 (1986). (c) King, R. B.: Polyhedron *1,* 132 (1982)
57. King, R. B.: J. Amer. Chem. Soc. *91,* 7211 (1969); *92,* 6455, 6460 (1970); *94,* 95 (1972)
58. Teo B. K.: Inorg. Chem. *23,* 1251 (1984); Teo, B. K., Longoni, G., Chung, F. R. K.: Inorg. Chem. *23,* 1258 (1984)
59. O'Neill, M. E., Wade, K.: Inorg. Chem. *21,* 461 (1982)
60. Pipal, J. R., Grimes, R. N.: Inorg. Chem. *18,* 257 (1979)
61. Bowser, J. R., Bonny, A., Pipal, J. R., Grimes, R. N.: J. Amer. Chem. Soc. *101,* 6229 (1979)
62. Mulliken, R. S.: J. Chem. Phys. *3,* 375 (1935)
63. Longuet-Higgins, H. C.: Quart. Rev. Chem. Soc. *11,* 121 (1957)
64. Longuet-Higgins, H. C., de V. Roberts, M.: Proc. Roy. Soc. A *230,* 110 (1955)
65. (a) Hoffmann, R., Lipscomb, W. N.: J. Chem. Phys. *36,* 2179 (1962). (b) Hoffmann, R., Lipscomb, W. N.: J. Chem. Phys. *36,* 3489 (1962); *37,* 2872 (1962)
66. Mulliken, R. S.: J. Chem. Phys. *23,* 1833, 2343 (1955)
67. Hoffmann, R.: J. Chem. Phys. *39,* 1397 (1963)
68. Klanberg, F., Muetterties, E. L.: Inorg. Chem. *5,* 1955 (1966); Eatin, D. R., Guggenberger, L. J., Muetterties, E. L.: Inorg. Chem. *6,* 1271 (1967); Guggenberger, L. J.: Inorg. Chem. *7,* 2260 (1968); Guggenberger L. J.: Inorg. Chem. *8,* 2771 (1969)
69. Dixon, D. A., Kleier, D. A., Halgren, T. A., Hall, J. H, Lipscomb, W. N.: J. Amer. Chem. Soc. *99,* 6226 (1977); Dewar, M. J. S., McKee, M. L.: Inorg. Chem. *17,* 1569 (1978); *19,* 2662 (1980); Guest, M. F., Hillier, I. H.: Mol. Phys. *26,* 435 (1973)
70. Williams, R. E.: Prog. Boron Chem. *2,* 51 (1970)
71. Williams, R. E.: Inorg. Chem. *10,* 210 (1971); Williams, R. E.: Adv. Inorg. Chem. Radiochem. *18,* 67 (1976)
72. (a) Wade, K.: J. Chem. Soc., Chem. Commun. 792 (1971). (b) See also: Wade, K.: Adv. Inorg. Chem. Radiochem. *18,* 1 (1976); Wade, K.: in Ref. 1b, Ch. 3 p. 193
73. Rudolph, R. W., Pretzer, W. R.: Inorg. Chem. *11,* 1974 (1972); Rudolph, R. W.: Acc. Chem. Res. *9,* 446 (1976)
74. Johnston, R. L., Mingos, D. M. P.: Inorg. Chem. *25,* 1661 (1986)
75. Albano, V., Chini, P., Scatturin, V.: J. Chem. Soc., Chem. Commun. 163 (1968)
76. Johnson, B. F. G., Johnston, R. D., Lewis, J.: J. Chem. Soc., Chem. Commun. 1057 (1967); Mason R., Robinson, W.: J. Chem. Soc., Chem. Commun. 468 (1968)
77. Albano, V., Bellon, P. L., Chini, P., Scatturin, V.: J. Organomet. Chem. *16,* 461 (1969)
78. Braterman, P. S.: Struct. Bond. *10,* 57 (1972)
79. Churchill, M. R., Wormwald, J., Knight, J., Mays, M. J.: J. Chem. Soc. Chem. Commun. 458 (1970)
80. Braye, E. H., Dahl, L. F., Hubel, W., Wampler, D. L.: J. Amer. Chem. Soc. *84,* 4633 (1962)
81. Wade, K.: Inorg. Nucl. Chem. Lett. *8,* 559, 563, 823 (1972); Wade, K.: Nature Phys. Sci. *240,* 71 (1972)
82. Mason, R., Thomas, K. M., Mingos, D. M. P.: J. Amer. Chem. Soc. *95,* 3802 (1973)
83. Evans, J.: J. Chem. Soc., Dalton Trans. 18, 35 (1978)
84. Mingos, D. M. P.: J. Chem. Soc., Dalton Trans. 133 (1974)
85. Lauher, J. W.: J. Amer. Chem. Soc. *100,* 5305 (1978)
86. Halpern, J.: Discuss. Faraday Soc. *46,* 7 (1968)
87. Foust, A. S., Foster, M. S., Dahl, L. F.: J. Amer. Chem. Soc. *91,* 5631 (1969)
88. Ellis, J. E.: J. Chem. Educ. *53,* 2 (1976)
89. Hertler, W. R., Klanberg, F., Muetterties, E. L.: Inorg. Chem. *6,* 1696 (1967)
90. Corbett, J. D.: Inorg. Chem. *7,* 198 (1968)
91. (a) Elian, M., Hoffmann, R.: Inorg. Chem. *14,* 1058 (1975). (b) Elian, M., Chen, M. M. L., Mingos, D. M. P., Hoffmann, R.: Inorg. Chem *15,* 1148 (1976)
92. Hoffmann, R.: Nobel Lecture 1981, Science *211,* 995 (1981); Hoffmann, R.: Angew. Chem. Int. Edn. Engl. *21,* 711 (1982)
93. Stone, F. G. A.: Angew. Chem. Int. Edn. Engl. *23,* 89 (1984) and references therein
94. DeKock, R. L., Fehlner, T. P.: Polyhedron *1,* 521 (1982)
95. Greenwood, N. N., Savory, C. G., Grimes, R. N., Sneddon, L. G., Davison, A., Wreford, S. S.: J. Chem. Soc., Chem. Commun. 718 (1974)

96. Miller, V. R., Weiss, R., Grimes, R. N.: J. Amer. Chem. Soc. *99,* 5646 (1977)
97. Brint, R. P., Pelin, K., Spalding, T. R.: Inorg. Nucl. Chem. Lett. *16,* 391 (1980); Brint, R. P., Spalding, T. R.: J. Chem. Soc., Dalton Trans. 1236 (1980)
98. Woolley, R. G.: Inorg. Chem. *24,* 3525 (1985)
99. Woolley, R. G.: Nouv. J. Chim. *5,* 219, 227, 441 (1981); Woolley, R. G.: in Ref. 1b, Ch. 9 p. 607; Woolley, R. G.: Inorg. Chem. *24,* 3519 (1985)
100. Nicholls, J. N.: Polyhedron *3,* 1307 (1984); Herrmann, W. A.: Angew. Chem. Int. Edn. Engl. *25,* 56 (1986) and references therein
101. Mingos, D. M. P., Forsyth, M. I.: J. Chem. Soc., Dalton Trans. 610 (1977)
102. Albano, V. G., Bellon, P. L., Ciani, G. F.: J. Chem. Soc., Chem. Commun. 1024 (1969)
103. Evans, D. G., Mingos, D. M. P.: Organometallics *2,* 435 (1983)
104. Goudsmit, R. J., Johnson, B. F. G., Lewis, J., Raithby, P. R., Whitmire, K. H.: J. Chem. Soc., Chem. Commun. 640 (1982)
105. McPartlin, M., Eady, C. R., Johnson, B. F. G., Lewis, J.: J. Chem. Soc., Chem. Commun. 883 (1976)
106. McPartlin, M.: Polyhedron *3,* 1279 (1984); McPartlin, M., Mingos, D. M. P.: Polyhedron *3,* 1321 (1984)
107. Evans, D. G., Mingos, D. M. P.: J. Organomet. Chem. *240,* 321 (1982)
108. Mingos, D. M. P.: J. Chem. Soc., Dalton Trans. 601 (1977)
109. Mingos, D. M. P.: J. Chem. Soc., Dalton Trans. 1163 (1976)
110. Nyholm, R. S.: Proc. Chem. Soc. Lond. 273 (1961); Pyykkö, P., Desclaux, J.-P.: Acc. Chem. Res. *12,* 276 (1979)
111. van Koten, G., Noltes, J. G.: in Comprehensive Organometallic Chemistry (eds. Wilkinson, G., Stone, F. G. A., Abel, E. W.,) Pergamon, Oxford 1982, Vol. 2 p. 709; Noltes, J. G.: Phil. Trans. R. Soc. Lond. *A308,* 35 (1982). For a theoretical treatment of bridged copper clusters see: Mehrotra, P. K., Hoffmann, R.: Inorg. Chem. *17,* 2187 (1978)
112. See, for example: Briant, C. E., Smith, R. G., Mingos, D. M. P.: J. Chem. Soc., Chem. Commun. 586 (1984)
113. Evans, D. G., Mingos, D. M. P.: J. Organomet. Chem. *232,* 171 (1982); *295,* 389 (1985)
114. Johnson, B. F. G., Kaner, D. A., Lewis, J., Raithby, P. R.: J. Organoment. Chem. *215,* C33 (1981); Johnson, B. F. G., Kaner, D. A., Lewis, J., Raithby, P. R., Taylor, M. J.: J. Chem. Soc., Chem. Commun. 314 (1982); Green, M., Orpen, A. G., Salter, I. D., Stone, F. G. A.: J. Chem. Soc., Chem. Commun. 813 (1982); Lehrer, H., Matt, D., Pregosin, P. S., Venanzi, L. M., Albinati, A.: J. Amer. Chem. Soc. *104,* 6825 (1982)
115. Mingos, D. M. P.: Pure Appl. Chem. *52,* 705 (1980)
116. Briant, C. E., Theobald, B. R. C., White, J. W., Bell, L. K., Mingos, D. M. P., Welch, A. J.: J. Chem. Soc., Chem. Commun. 201 (1981)
117. Mingos, D. M. P.: Phil. Trans. R. Soc. Lond. *A308,* 75 (1982); Hall, K. P., Gilmour, D. I., Mingos, D. M. P.: J. Organomet. Chem. *268,* 275 (1984); Mingos, D. M. P.: Polyhedron *3,* 1289 (1984)
118. Mingos, D. M. P.: J. Chem. Soc., Chem. Commun. 706 (1983)
119. Mingos, D. M. P.: Acc. Chem. Res. *17,* 311 (1984)
120. Mingos, D. M. P., Evans, D. G.: J. Organomet. Chem. *251,* C13 (1983)
121. Chini, P.: J. Organomet. Chem. *200,* 37 (1980)
122. Lauher, J. W.: J. Amer. Chem. Soc. *101,* 2604 (1979)
123. Ciani, G., Sironi, A.: J. Organomet. Chem. *197,* 233 (1980)
124. Slovokhotov, Yu. L., Struchkov, Yu. T.: Russ. Chem. Rev., *54,* 323 (1985)
125. Hume-Rothery, W.: The Metallic State, Oxford University Press, New York, 1931, p. 328
126. Teo, B. K.: J. Chem. Soc., Chem. Commun. 1362 (1983)
127. (a) Messmer, R. P., Knudson, S. K., Johnson, K. H., Diamond, J. B., Yang, C. Y.: Phys. Rev. B: Solid State *13,* 1396 (1976). (b) Yang, C. Y., Johnson, K. H., Salahub, D. R., Kaspar, J., Messmer, R. P.: Phys. Rev. B: Solid State *24,* 5673 (1981)
128. (a) Mingos, D. M. P.: J. Chem. Soc., Chem. Commun. 1352 (1985). (b) Mingos, D. M. P.: Chem. Soc. Rev., *15,* 31 (1986)
129. Vollenbroek, F. A., van der Berg, J. P., van der Velden, J. W. A., Bour, J. J.: Inorg. Chem. *19,* 2685 (1980)
130. Burdett, J. K.: Prog. Solid State Chem. *15,* 173 (1985)

131. Mingos, D. M. P.: Anglo-Japanese Advanced Research Meeting: Molecular Basis of Transition Metal Catalysis, April 1985, Abstracts, p. 9
132. Shustorovich, E.: N.A.T.O. A.S.I. Ser., Quantum Chemistry: The Challenge of Transition Metals and Coordination Chemistry (Ed. Veillard, A.,) Reidel, Dordrecht (NL) 1986, 445; Shustorovich, E.: Surf. Sci. Rep. *6,* 1 (1986)
133. Benfield, R. E., Edwards, P. P., Stacy, A. M.: J. Chem. Soc., Chem. Commun. 525 (1982); Benfield, R. E., Edwards, P. P., Nelson, W. J. H., Vargas, M. D., Johnson, D. C., Sienko, M. J.: Nature *314,* 231 (1985); Teo, B. K., DiSalvo, F. J., Waszczak, J. V., Longoni, G., Ceriotti, A.: Inorg. Chem. *25,* 2262 (1986)
134. Benfield, R. E.: unpublished work, mentioned at the Dalton/A.C.S. meeting: Clusters in the Solid State, Birmingham (U.K.) 1986
135. Pimental, G. C., Pitzer, K. S.: J. Chem. Phys. *17,* 882 (1949)
136. See, for example: DeKock, R. L., Gray, H. B.: Chemical Structure and Bonding, Benjamin/Cummings, Menlo Park (California) 1980, p. 419
137. Bayliss, N. S.: J. Chem. Phys. *16,* 287 (1948); Kuhn, H.: J. Chem. Phys. *16,* 840 (1948); Platt, J. R.: J. Chem. Phys. *22,* 1448 (1954); Wheland, G. R.: Resonance in Organic Chemistry, Wiley, New York 1955
138. Lu, C.-S., Donohue, J.: J. Amer. Chem. Soc. *66,* 818 (1944)
139. Chapman, D., Waddington, T. C.: Trans. Faraday Soc. *58,* 1679 (1962)
140. Hoffmann, R., Gouterman, M.: J. Chem. Phys. *36,* 2189 (1962)
141. Bethe, H. A.: Ann. Physic. *3,* 133 (1929); Van Vleck, J. H.: J. Chem. Phys. *3,* 807 (1935); Griffith, J. S.: The Theory of Transition Metal Ions, Cambridge University Press, Cambridge 1961; Ballhausen, C. J.: Ligand Field Theory, McGraw-Hill, New York 1962
142. Waddington, T. C.: Trans. Faraday Soc. *63,* 1313 (1967)
143. Hoffmann, D. K., Ruedenberg, K., Verkade, J. G.: Struct. Bond. *33,* 57 (1977)
144. Mulliken, R. S.: Rev. Mod. Phys. *2,* 60, 506 (1930); *4,* 40 (1932)
145. (a) Stone, A. J.: Mol. Phys. *41,* 1339 (1980). (b) Stone, A. J.: Inorg. Chem. *20,* 563 (1981). (c) Stone, A. J., Alderton, M. J.: Inorg. Chem. *21,* 2297 (1982). (d) Stone, A. J.: Polyhedron *3,* 2051 (1984)
146. Brink, D. M., Satchler, G. R.: Angular Momentum (2nd. Edn.), Oxford University Press, Oxford 1971
147. Eckart, C.: Phys. Rev. *36,* 878 (1930)
148. Hückel, E.: Z. Physik. *70,* 204 (1931); *72,* 310 (1931); *76,* 628 (1932)
149. Brint, P., Cronin, J. P., Seward, E.: J. Chem. Soc., Dalton Trans. 675 (1983)
150. Fowler, P. W., Porterfield, W. W.: Inorg. Chem. *24,* 3511 (1985)
151. Johnston, R. L., Mingos, D. M. P.: J. Organomet. Chem. *280,* 419 (1985)
152. Johnston, R. L., Mingos, D. M. P.: J. Organomet. Chem. *280,* 407 (1985)
153. Biggs, N. L.: Algebraic Graph Theory, Cambridge University Press, London 1974
154. Redmond, D. B., Quinn, C. M., McKiernan, J. G. R.: J. Chem. Soc., Faraday Trans. II *79,* 1791 (1983); Inorg. Chem. *22,* 2310 (1983); J. Chem. Educ. *61,* 569, 572 (1984)
155. McDonnell Jr., P. W.: Introduction to Map Making, Marcell Dekker, New York 1979
156. (a) Fowler, P. W.: Polyhedron *4,* 2051 (1985). (b) Fowler, P. W., Woolrich, J.: Chem. Phys. Lett. *127,* 84 (1986)
157. (a) Johnston, R. L., Mingos, D. M. P.: J. Chem. Soc., Dalton Trans. 647 (1987). (b) Johnston, R. L., Mingos, D. M. P.: J. Chem. Soc., Dalton Trans., 1445 (1981)
158. Altmann, S. L.: Induced Representations in Crystals and Molecules, Academic Press, New York 1977
159. Ceulemans, A.: Mol. Phys. *54,* 161 (1985)
160. See, for example: Cotton, F. A.: Chemical Applications of Group Theory (2nd Edn.), Wiley, New York 1971
161. Maier, G., Pfriem, S., Schafter, U., Matusch, R.: Angew. Chem. *90,* 652 (1978)
162. Atoji, M., Lipscomb, W. N.: Acta Crystallogr. *6,* 547 (1953)
163. Brown, L. D., Lipscomb, W. N.: Inorg. Chem. *16,* 2989 (1977); Bicerano, G., Marynick, D. S., Lipscomb, W. N.: Inorg. Chem. *17,* 2041, 3443 (1978)
164. Johnston, R. L., Mingos, D. M. P.: Polyhedron *5,* 2059 (1986)
165. Bould, J., Greenwood, N. N., Kennedy, J. D., McDonald, W. S.: J. Chem. Soc., Chem. Commun. 465 (1982). For further examples of *hypercloso* (or *iso-closo*) metallaboranes see: Baker, R. T.: Inorg. Chem. *25,* 109 (1986); Kennedy, J. D.: Inorg. Chem. *25,* 111 (1986);

Kennedy, J. D.: Prog. Inorg. Chem. *32,* 519 (1984); *34,* 211 (1986); Greenwood, N. N.: Chem. Soc. Rev. *13,* 353 (1984) and references therein
For a discussion of the bonding in these molecules see: Johnston, R. L., Mingos, D. M. P.: Inorg. Chem. *25,* 3321 (1986); Johnston, R. L.: D. Phil. Thesis, University of Oxford 1986
166. Hoffmann, R., Schilling, B. E. R., Bau, R., Kaesz, H. D., Mingos, D. M. P.: J. Amer. Chem. Soc. *100,* 6088 (1978); Schilling, B. E. R., Hoffmann, R.: J. Amer. Chem. Soc. *101,* 3456 (1979)
167. (a) Bullett, D. W.: Phys. Rev. Lett. *39,* 664 (1977); Chang, K. W., Woolley, R. G.: J. Phys. C *12,* 2745 (1979). (b) Seifert, G., Mrosan, E., Müller, H.: Phys. Status Solidi B *89,* 553 (1978); Cotton, F. A., Stanley, G. G.: Chem. Phys. Lett. *58,* 450 (1978); Johnson, K. H.: Crit. Rev. Solid State Mater. Sci. *7,* 101 (1978); Messmer, R. P.: Surf. Sci. *106,* 225 (1981); Case, D. A.: Annu. Rev. Phys. Chem. *33,* 151 (1982)
168. Demuynck, J., Rohmer, M.-M., Strich, A., Veillard, A.: J. Chem. Phys. *75,* 3443 (1981)
169. Hall, M. B., Fenske, R. F.: Inorg. Chem. *11,* 768 (1972); Teo, B. K., Hall, M. B., Fenske, R. F., Dahl, L. F.: J. Organomet. Chem. *70,* 413 (1974); Cheskey, P. T., Hall, M. B.: Inorg. Chem. *20,* 4419 (1981); Sherwood Jr., D. E., Hall, M. B.: Inorg. Chem. *21,* 3458 (1982); Sherwood Jr., D. E., Hall, M. B.: Organometallics *1,* 1519 (1982)
170. See, for example: Korol'kov, D. V., Miessner, H.: Z. Phys. Chem. *253,* 25 (1973)
171. Manning, M. C., Trogler, W. C.: Coord. Chem. Rev. *38,* 89 (1981)
172. Andersen, O. K., Woolley, R. G.: Mol. Phys. *26,* 905 (1973); Andersen, O. K.: Phys. Rev. B: Solid State *12,* 3060 (1975); Andersen, O. K., Klose, W., Nohl, H.: Phys. Rev. B: Solid State *17,* 1209 (1978)
173. See, for example: Albright, T. A., Burdett, J. K., Whangbo, M.-H.: Orbital Interactions in Chemistry, Wiley, New York 1984, Ch. 13 p. 229; Hoffmann, R., Zheng, C.: N.A.T.O. A.S.I. Ser., Quantum Chemistry: The Challenge of Transition Metals and Coordination Chemistry (Ed. Veillard, A.), Reidel, Dordrecht (NL) 1986, p. 425; Burdett, J. K., Canadell, E., Hughbanks, T.: J. Amer. Chem. Soc. *108,* 3971 (1986); Burdett, J. K.: Struct. Bond., *65,* 29 (1987)
174. Wei, C.-H., Dahl, L. F.: J. Amer. Chem. Soc. *91,* 1351 (1969)
175. Benfield, R. E., Johnson, B. F. G.: Top. Stereochem. *12,* 253 (1981) and references therein
176. Kepert, D. L., Williams, S. C.: J. Organomet. Chem. *217,* 235 (1981)
177. Mingos, D. M. P.: Inorg. Chem. *21,* 464 (1982)
178. Lauher, J. W.: J. Amer. Chem. Soc. *108,* 1521 (1986)
179. Clare, B. W., Favas, M. C., Kepert, D. L., May, A. S.: Adv. Dynamic Stereochem. *1,* 1 (1985) and references therein
180. Berkert, U., Allinger, N. L.: Molecular Mechanics, American Chemical Society, Washington 1982
181. Evans, D. G.: D. Phil. Thesis, University of Oxford 1984 and references therein
182. Evans, D. G.: J. Chem. Soc., Chem. Commun. 675 (1983)
183. Minot, C., Criado-Sancho, M.: Nouv. J. Chim. *8,* 537 (1984)

Magnetism of the Heteropolymetallic Systems

Olivier Kahn

Laboratoire de Spectrochimie des Eléments de Transition, Unité Associée au CNRS n° 420, Université de Paris-Sud, 91 405 Orsay, France

The field of heteropolymetallic systems with magnetic metal centers occupies a crossing point between biology and physics. For instance the Cu(II)–Fe(III) interaction in cytochrome oxidase is of the same nature as the Cu(II)–Mn(II) interaction in a novel system which could be the first molecular ferromagnet. The mechanism of the interaction is discussed, both from a phenomenological view point using a spin Hamiltonian, and from an orbital view point. An orbital model for the isotropic interaction is presented. It is based on the concept of natural magnetic orbitals. The mechanism of the anisotropic and antisymmetric interactions is more briefly treated. The role of the Zeeman perturbation is then considered in relation with the magnetic and EPR properties of the heterobimetallic complexes. Several examples are presented to emphasize that the nature, ferro- or antiferromagnetic of the isotropic interaction is controlled by the symmetry of the magnetic orbitals. The concept of overlap density is introduced. It permits an estimation of the magnitude of the ferromagnetic stabilization in the case of orthogonality of the magnetic orbitals. The Cu(II)–Fe(III) interaction, in relation to the situation encountered in cytochrome oxidase, the Cu(II)–Ni(II) interaction and a few additional selected examples are discussed. A section deals with the case where one of the interacting ions has an orbital degeneracy. Afterwards, the heterotrinuclear complexes are studied. The important concept of regular and irregular spin state structure is developped and the Mn(II)Cu(II)Mn(II) triad is presented as a spectacular example of irregular spin state structure. A section is devoted to the ordered bimetallic chains. The theory is presented, both at a qualitative and quantitative levels and the already reported compounds of this kind are discussed. One of them may be considered as one of the first molecular ferromagnets. The last but one section concerns the systems with even more subtle spin orders. In conclusion, the vast perspectives of this area are outlined.

Structure and Bonding, Vol. 68

1 Introduction

Apparently nature has for ever utilized heteropolymetallic sites to realize some biological processes. The copper-iron site in cytochrome oxidase and the copper-zinc site in superoxide dismutase are two well known examples of this. Maybe nature has chosen heteropolymetallic sites when the effect desired was so subtle that no homometallic site could achieve it.

For more than a decade, the design of heteropolymetallic complexes has also been a challenging field for the inorganic chemists, in which they can make use of their promethean power to transform matter, either to mimic what nature has done for thousands of years, or to obtain systems exhibiting quite novel properties. In general such a design requires either the use of dissymmetric polynucleating ligands[1–5], or to work incrementally by using the strategy of "complexes as ligands"[6–13]. Several families of complexes of this kind are now known. In this review, we are essentially interested in those in which the metal centers are magnetic. In this case, the complexes exhibit magnetic properties which are not simply the sum of those of the individual ions surrounded by their nearest neighbor ligands. These properties result from both the nature and the magnitude of the interactions between the metal ions within the molecular unit. Accurate magnetic and EPR data on heteropolymetallic complexes are much less numerous than those dealing with homopolymetallic compounds, particularly copper(II) compounds. However, it is likely not an exaggeration to claim that some of these data opened quite new perspectives in the field of the molecular magnetism. More recently, novel types of heteropolymetallic systems, like ordered bimetallic chains or other one-dimension compounds with even more subtle spin orders have been described. Some of these systems exhibit magnetic behaviors previously unknown. The concept of subtle spin order could lead to the first genuine molecular ferromagnets.

As we will see, the heteropolymetallic systems are of interest for both biologists and bioinorganic chemists investigating the structure and the role of the polymetallic active sites in biological processes, and the physicists or the physical inorganic chemists seeking to design new magnetic materials. In fact, it is rather fascinating to realize that the interaction between copper(II) and iron(III) (d^9 and d^5 ions respectively) which occurs in cytochrome oxidase, a metalloenzyme of which the structure is still a mystery, is exactly of the same physical nature as the interaction between copper(II) and manganese(II) in a material which may well be a good model for a molecular ferromagnet.

When a field occupies such a crossing point between biology and the physics, it deserves to be explored in a thorough manner. This is the goal of this article.

I do not intend at all the present a comprehensive introduction of the field. On the contrary, I shall select some problems that will be discussed in more detail. To some extent, the choice is a reflection of my own interests. It is largely based on the idea that the concept is more important than the mathematical technique and that phenomena involving energies of several tens of wavenumbers deserve a greater emphasis than the phenomena of very weak magnitude. It is also my opinion that what is important is the compound itself. The physical methods are tools to investigate the compounds and not goals in themselves. In writing this article, I have also in mind that the eventual applications, far from being a brake for the fundamental research, can be a source of inspiration.

This explains why a significant part of this work will be devoted to the design of ferromagnetically coupled systems and of molecular ferromagnets, which is one of the main challenges in the field of the molecular materials.

2 Mechanisms of the Interactions Between Two Orbitally Non Degenerate Metal Ions

In this section, I shall present some key concepts which facilitate the understanding of the mechanisms of the interactions between two metal ions of which the ground state is an orbital singlet. Much later (Sect. 8), we shall approach the much more complicated case where one of the interacting ions has an orbital degeneracy at the first order.

2.1 Phenomenological Approach

Let us consider an A − B heterobimetallic complex. A and B represent magnetic metal ions metal ions surrounded by their ligands. In other words, A and B may be considered as monomeric fragments and have in common one or more bridging ligands. We have n_A unpaired electrons around A of which the local ground state is $^{2S_A+1}\Gamma_A$ and n_B unpaired electrons around B with a $^{2S_B+1}\Gamma_B$ local ground state. Γ_A and Γ_B note the orbital symmetries of the local ground states. In other words, Γ_A and Γ_B are irreducible representations of the site groups. We have of course $S_A = n_A/2$ and $S_B = n_B/2$. If S_A (or S_B) is larger than 1/2, the local ground state may be split in zero field owing to the synergetic effect of the second order spin-orbit coupling and the local distortion. This effect is often called the local anisotropy and is phenomenologically described by:

$$\begin{aligned} \mathcal{H}_A &= \hat{S}_A \cdot \mathbf{D}_A \cdot \hat{S}_A \\ \mathcal{H}_B &= \hat{S}_B \cdot \mathbf{D}_B \cdot \hat{S}_B \end{aligned} \qquad (1)$$

where $\mathbf{D}_A$ and $\mathbf{D}_B$ are symmetric and traceless tensors. The interaction within A − B may be decomposed into three terms: (i) the isotropic interaction which in most of the cases is the dominant term. It leads to low lying pair states characterized by S varying by integer value from $|S_A - S_B|$ to $S_A + S_B$. If the whole symmetry of A − B and the site symmetries of the A and B fragments are identical, the low lying pair states will be labelled as $^{2S+1}\Gamma$ with Γ given by the direct product $\Gamma_A \times \Gamma_B$. This isotropic interaction is phenomenologically described by $-J\hat{S}_A \cdot \hat{S}_B$ where J is a scalar. When J is negative, the interaction is antiferromagnetic and the ground state has the lowest spin multiplicity $|S_A - S_B|$. When J is positive, the interaction is ferromagnetic and the ground state has the highest spin multiplicity $S_A + S_B$; (ii) the anisotropic interaction which splits in zero field the $^{2S+1}\Gamma$ pair states with $S > 1/2$. This anisotropic interaction may be phenomenologically described by $\hat{S}_A \cdot \mathbf{D}_{AB} \cdot \hat{S}_B$ where $\mathbf{D}_{AB}$ is again a symmetric and traceless tensor; (iii) the antisymmetric interaction finally which gives an additional splitting in zero field of the pair states and mixes the components of same M_S belonging to different states. This antisymmetric interaction, of which the phenomenological expression is $\vec{d} \cdot \hat{S}_A \wedge \hat{S}_B$, $\vec{d}$ being a vector, vanishes when A-B has a C_{nv} symmetry with $n \geq 2$[14]. Many of the complexes presented in the following sections have a symmetry very close to C_{2v}, consequently d will approach 0.

All together the phenomenological Hamiltonian appropriate for describing the low lying levels of A − B in zero field is:

$$\mathcal{H}_{AB} = -J\hat{S}_A \cdot \hat{S}_B + \hat{S}_A \cdot \mathbf{D}_A \cdot \hat{S}_A + \hat{S}_B \cdot \mathbf{D}_B \cdot \hat{S}_B + \hat{S}_A \cdot \mathbf{D}_{AB} \cdot \hat{S}_B + \vec{d} \cdot \hat{S}_A \wedge \hat{S}_B \quad (2)$$

It should be noticed here that the local anisotropies as well as the anisotropic and antisymmetric interactions contribute to the splitting of the $^{2S+1}\Gamma$ states and experimentally it is not possible to distinguish between these contributions. There is even an additional origin to the zero field splitting of the pair states, namely the interaction between the magnetic dipoles located in A and B respectively. This dipolar magnetic interaction can be calculated in principle either by assuming that the dipoles are centered on the metal ions (point approximation), or by taking into account the delocalization of the spin densities in the A and B fragments as given by the expressions of the magnetic orbitals (see Sect. 2.2).

In Eq. (2), I did not include smaller effects like hyperfine structure terms. In this article, we are more concerned with pure compounds in the solid state than with dilute species either in solution or in host lattices, so that the hyperfine structure is generally not resolved in EPR. For completeness, however, we give the additional terms in (3) appropriate to describe the hyperfine structure of A − B:

$$\mathcal{H}_{HS} = \hat{I}_A \cdot \mathbf{A}_A \cdot \hat{S}_A + \hat{I}_B \cdot \mathbf{A}_B \cdot \hat{S}_B \quad (3)$$

$\mathbf{A}_A$ and $\mathbf{A}_B$ are the local hyperfine tensors and $\hat{I}_A$ and $\hat{I}_B$ the nuclear spin operators. In Eq. (3), superhyperfine and transferred hyperfine interactions are neglected.

The Hamiltonian (2) or (2) + (3) is relatively easy to handle, but it does not provide any information on the real mechanisms of the interactions. It has absolutely no predictive character. The crucial problem is then to find the relations between the parameters appearing in Eq. (2) and the molecular integrals relating to the orbital nature of the phenomena.

2.2 Concept of Natural Magnetic Orbitals

So far two types of model have been proposed to describe the isotropic interaction between magnetic centers[15]. The basic difference between these models is the nature of the magnetic orbitals they use.

The first model is due to Anderson and its original formulation dealt with extended lattices like oxides or fluorides[16]. Its molecular variant has been introduced by Hoffmann et al.[17]. This model uses orthogonalized magnetic orbitals obtained by localization of the singly occupied Self Consistent Field (S.C.F.) molecular orbitals in the state of highest spin multiplicity. The procedure of localization is well established for the symmetric systems in which the magnetic orbitals are symmetry related. On the other hand, the situation is much more complicated for the dissymmetric systems like the heteropolymetallic complexes and to the best of our knowledge, there has been no attempt to adapt Anderson's model to this kind of compounds.

The alternative model uses natural magnetic orbitals[18–21)] and can be easily adapted to non symmetric homobinuclear and heterobinuclear compounds. The basic idea of natural magnetic orbitals directly derives from the description of A – B from the A and B monomeric fragments with a common zone corresponding to the bridges. The n_A unpaired electrons around A are described by n_A orbitals $\phi_{\mu A}$ that are the singly occupied molecular orbitals for the A fragment in its local ground state $^{2S_A+1}\Gamma_A$. In the same way the n_B unpaired electrons around B are described by the $\phi_{\nu B}$ magnetic orbitals. One sees that $\phi_{\mu A}$ and $\phi_{\nu B}$ are one-electron eigenfunctions of local Hamiltonians. However, in many cases, it is also possible to give them a more chemical meaning. Let me give an example of this. In Sect. 6.2, we shall study in detail a copper(II) nickel(II) complex noted [CuNi] of the type shown below, with A = Cu(II) in planar surroundings and

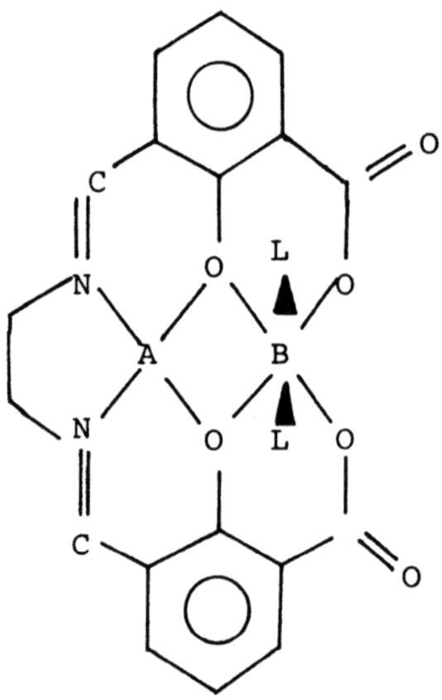

L = H_2O

B = Ni(II) in octahedral surroundings. We have $n_A = 1$ magnetic orbital around copper and $n_B = 2$ magnetic orbitals around nickel. Two other complexes of the same series are known, namely [CuMg] and [NiNi], that are isostructural with [CuNi]. In the former, Mg(II) is diamagnetic and there is just one unpaired electron which may be considered as the ϕ_{1Cu} magnetic orbital of [CuNi]. In the latter, the Ni(II) in the inside site is also diamagnetic while the second Ni(II) possesses two unpaired electrons occupying singly occupied orbitals which may be considered as the ϕ_{1Ni} and ϕ_{2Ni} magnetic orbitals of [CuNi].

2.3 Isotropic Interaction

If we assume that only the $N = n_A + n_B$ unpaired electrons are active in the isotropic interaction phenomenon, then the electrostatic Hamiltonian suitable for the problem may be written as:

$$\mathcal{H} = \sum_{i=1}^{N} h(i) + \sum_{i=1}^{N} \sum_{j>i}^{N} \frac{1}{r_{ij}} \tag{4}$$

h(i) takes into account the kinetic energy of the electron i and the one-electron potential acting on it, due to the nuclei and the passive electrons; r_{ij} is the interelectronic distance. The isotropic interaction is purely orbital in nature and corresponds to the extreme case of weak bond between the A and B fragments, such that the first excited states in A – B are low enough in energy to be thermally populated. It follows that it is appropriate to give the form of Heitler-London functions to the eigenfunctions associated with those low lying states. It is sufficient to work in the subspace $M_S = 0$ if N is even or $M_S = 1/2$ if N is odd since all the pair states will have such a component with $M_S = 0$ or 1/2. The basis set is then made out of all the Heitler-London functions with $M_S = 0$ (1/2) built from the n_A magnetic orbitals $\phi_{\mu A}$ and n_B magnetic orbitals $\phi_{\nu B}$ (noted μ_A and ν_B in the remainder of this section to simplify the notation). Since the magnetic orbitals belonging to different monomeric fragments can, in principle, overlap ($\langle \mu_A/\nu_B \rangle \neq 0$), the Heitler-London functions are neither normalized nor orthogonal to each other. We do not develop the calculation which is rather tedious but without conceptual difficulty. The method can be found in earlier reference[22, 23]. It consists of expanding the energies of the low lying pair states according to the increasing powers of the overlap integrals $S_{\mu\nu} = \langle \mu_A/\nu_B \rangle$. If $S_{\mu\nu}^2$ is small with respect to 1, then the relative energies of the pair states are found to follow the Lande interval rule:

$$E(S-1) - E(S) = JS \tag{5}$$

to which the spin Hamiltonian $-J\hat{S}_A \cdot \hat{S}_B$ leads. J can then be related to the integrals involving the magnetic orbitals according to:

$$J = \frac{1}{n_A n_B} \sum_{\mu=1}^{n_A} \sum_{\nu=1}^{n_B} J_{\mu\nu} \tag{6}$$

The expression for $J_{\mu\nu}$ limited to the zeroth and first order terms in $S_{\mu\nu}$ is:

$$J_{\mu\nu} = 2\, j_{\mu\nu} + 4\, t_{\mu\nu}S_{\mu\nu} + 4 \left\{ \sum_{\substack{\omega=1 \\ \neq\mu}}^{n_A} [(\omega_A\omega_A|\mu_A\nu_B) - (\omega_A\mu_A|\omega_A\nu_B)] + \sum_{\substack{\varrho=1 \\ \neq\nu}}^{n_B} [(\mu_A\nu_B|\varrho_B\varrho_B) - (\mu_A\varrho_B|\varrho_B\nu_B)] \right\} S_{\mu\nu} \tag{7}$$

where $j_{\mu\nu}$ is the two-electron exchange integral:

$$j_{\mu\nu} = \left\langle \mu_A(i)\nu_B(j) \left| \frac{1}{r_{ij}} \right| \mu_A(j)\nu_B(i) \right\rangle \tag{8}$$

and $t_{\mu\nu}$ is the one-electron transfer integral:

$$t_{\mu\nu} = \left\langle \mu_A(i) | h(i) - \frac{\alpha_{\mu A} + \alpha_{\nu B}}{2} | \nu_B(i) \right\rangle \tag{9}$$

with:

$$\alpha_{\mu A} = \langle \mu_A(i)|h(i)|\mu_A(i)\rangle \tag{10}$$

$(\mu_A\nu_B|\omega_C\varrho_D)$ is a simplified notation for the two-electron integral $\langle\mu_A(i)\omega_C(j)|1/r_{ij}|\nu_B(i)\varrho_D(j)\rangle$. The first term in the right hand of Eq. (7), $2\ j_{\mu\nu}$, is always positive and favors a ferromagnetic interaction. When the two magnetic orbitals μ_A and ν_B are orthogonal with $S_{\mu\nu} = 0$, $J_{\mu\nu}$ reduces to this ferromagnetic contribution. The second term, $4\ t_{\mu\nu}S_{\mu\nu}$, for $S_{\mu\nu}$ not too large, is negative and favors an antiferromagnetic interaction. In the Mulliken approximation, $t_{\mu\nu}$ is expressed as:

$$t_{\mu\nu} = KS_{\mu\nu}\frac{\alpha_{\mu A} + \alpha_{\nu B}}{2} \tag{11}$$

so that this antiferromagnetic contribution varies roughly as $-S_{\mu\nu}^2$. As for the term in $S_{\mu\nu}$ involving two-electron integrals, it is difficult to estimate its importance or even its sign. The integrals of the type $(\omega_A\omega_A|\mu_A\nu_B)$ appear larger than those of the type $(\omega_A\mu_A|\omega_A\nu_B)$. If so, this term would be positive. However, it is important to realize that this term does not exist when there is just one unpaired electron per magnetic center with a singlet state lower or higher in energy than a triplet state according to the nature antiferro- or ferromagnetic of the interaction. Therefore, it does not play the key role for the nature of the phenomenon.

To sum up, the sign of $J_{\mu\nu}$ results from a competition between two key contributions, the ferromagnetic and the antiferromagnetic ones; the former varies as the two-electron exchange integral $j_{\mu\nu}$ and takes over when the overlap integral $S_{\mu\nu}$ is zero or close to zero. The latter varies as $-S_{\mu\nu}^2$ and takes over as soon as $S_{\mu\nu}$ is no longer very small. In Sect. 4, we shall see how this image of competition between ferro- and antiferromagnetic contributions is heuristic and allows to predict both the nature and the order of magnitude of the isotropic interaction.

If the A and B fragments have the same site symmetry and that around B with $n_B \geqslant n_A$, there is never more than one magnetic orbital μ_B giving a non zero overlap with the magnetic orbital μ_A, then Eq. (7) becomes:

$$J_{\mu\nu} = 2\ j_{\mu\nu} + 4\ t_{\mu\mu}S_{\mu\mu}\delta_{\mu\nu} + 4\left\{\sum_{\substack{\omega=1\\ \neq\mu}}^{n_A}[(\omega_A\omega_A|\mu_A\mu_B) - (\omega_A\mu_A|\omega_A\mu_B)] + \sum_{\substack{\varrho=1\\ \neq\mu}}^{n_B}[(\mu_A\mu_B|\varrho_B\varrho_B) - (\mu_A\varrho_B|\varrho_B\mu_B)]\right\}S_{\mu\mu}\delta_{\mu\nu} \tag{12}$$

In this equation, μ denotes the irreducible representation spanned by μ_A and μ_B. From each pair of magnetic orbitals μ_A and μ_B localized on the A and B fragments respectively, one can construct two molecular orbitals delocalized on the whole molecule A – B. Each of these molecular orbitals is singly occupied in the pair state of highest spin multiplicity. We denote $\Delta_{\mu\mu}$ the energy gap between these molecular orbitals for this state and $D_{\mu\mu}$ the energy gap between the two magnetic orbitals as shown below.

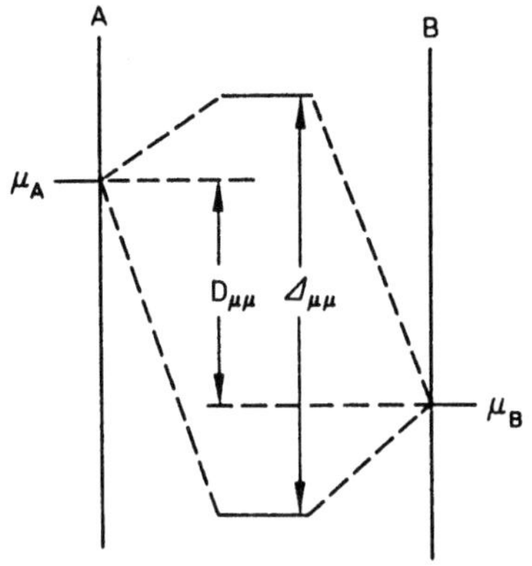

Using the method given in Refs. 22 and 23, Eq. (12) may be reexpressed as[24]:

$$J_{\mu\nu} = 2\, j_{\mu\nu} - 2\,(\Delta^2_{\mu\mu} - D^2_{\mu\mu})^{1/2} S_{\mu\mu}\delta_{\mu\nu} \tag{13}$$

In this equation, two terms again appear with the former favoring the ferromagnetic and the latter favoring the antiferromagnetic interactions. This expression (13) clearly points out that the antiferromagnetic contribution corresponds to the tendency for two electrons originally present in two magnetic orbitals of the same symmetry to pair in a molecular orbital of low energy. This trend is governed by the overlap between the magnetic orbitals. When this overlap is zero, due to the orthogonality of the magnetic orbitals, the parallel-spin situation is favored, as expected from Hund.

2.4 Anisotropic and Antisymmetric Interactions

In the last section, we did not take into account spin-orbit coupling. This is only valid if all the local excited states are much too high in energy to interact with the local ground states. Actually, in each monomeric fragment, A or B, the ground state can couple with the excited states through spin-orbit coupling. This coupling leads not only to the local anisotropies defined in Eq. (1) but also the anisotropic and antisymmetric exchange interactions within A – B[14, 25, 26].

The expressions of the components of the $\mathbf{D}_{AB}$ tensor and of the $\vec{d}$ vector appearing in Eq. (2) might, in principle, be derived in the framework of the formalism used in the previous section. For that, it is necessary to replace the zeroth order magnetic orbitals defined in Sect. 2.2 by new functions taking into account the spin-orbit perturbation up to the second order, then on the basis set of the new Heitler-London functions to diagonalize the electrostatic Hamiltonian (4). In the general case this calculation is extremely tedious. For instance, it is no longer possible to work only in the $M_S = 0$ (or 1/2) subspace since $\vec{d} \cdot \hat{S}_A \wedge \hat{S}_B$ mixes the pair states. So far only one specific case has been treated, with $S_A = S_B = 1/2$[15]. Kanamori[26, 27], however, gave the following general expressions for the principal values D_{uu} of $\mathbf{D}_{AB}$ and the components d_u of $\vec{d}$, u = x, y, z:

$$D_{uu} = D^A_{uu} + D^B_{uu} \tag{14}$$

$$d_u = d_u^A - d_u^B \tag{15}$$

$$D_{uu}^A = \sum_e \frac{\lambda_A^2}{\Delta_{e_A}^2} \langle g_A|\hat{L}_u^A|e_A\rangle \langle e_A|\hat{L}_u^A|g_A\rangle \, J^{(e_A g_B e_A g_B)} \tag{16}$$

$$d_u^A = 2\sum_e \frac{\lambda_A}{\Delta_{e_A}} \langle g_A|\hat{L}_u^A|e_A\rangle \, J^{(e_A g_B g_A g_B)} \tag{17}$$

g_A specifies the ground state of A with e_A one of the excited states located at an energy Δ_{e_A} above $g_A \cdot \hat{L}_u^A$ is a component of the orbital momentum, λ_A the local spin orbit coupling parameter and $J^{(k_A l_B m_A n_B)}$ is an exchange integral of the form $\langle k_A l_B|\mathcal{H}|m_A n_B\rangle$.

The most important point emerging from Eqs. (16) and (17) is that the anisotropic and antisymmetric interactions are not directly related to the isotropic interaction. Whereas the latter only involves the two local ground states, the former involve both local ground and excited states. More precisely the anisotropic interaction arises form the synergistic effect of the local spin-orbit perturbations and the interaction between the ground state of one fragment and the excited states of the other. A last point deserves to be underlined. In an A − B system, the isotropic interaction is most often antiferromagnetic owing to the overlap between the magnetic orbitals. On the other hand, if the symmetry is high enough, the ground state of one ion may be orthogonal to all the excited states of the other so that the interactions involved in the anisotropic term would be ferromagnetic[28-30].

3 Zeeman Perturbation, Magnetic Susceptibility and EPR Spectra

Up to now, we have considered the A − B system in absence of any external perturbation. However the most appropriate methods to study the nature and the relative energies of the low lying states in A − B consist of observing the response of the system to an external magnetic field $\vec{H}$. In this case, we have to add to Eq. (2) [or (2) + (3)] a Zeeman perturbation of the form:

$$\mathcal{H}_Z = \beta\vec{H} \cdot (\mathbf{g}_A \cdot \hat{S}_A + \mathbf{g}_B \cdot \hat{S}_B) \tag{18}$$

where $\mathbf{g}_A$ and $\mathbf{g}_B$ are the local g-tensors and β the Bohr magneton. When, as in most of the examples discussed in this article, the isotropic interaction is by far the most important term in the Hamiltonian, including the Zeeman perturbation, it is possible to associate an effective spin Hamiltonian $\mathcal{H}_S$ with each of the $^{2S+1}\Gamma$ pair states:

$$\mathcal{H}_S = \hat{S} \cdot \mathbf{D} \cdot \hat{S} + \beta\vec{H} \cdot \mathbf{g}_S \cdot \hat{S} + \hat{I}_A \cdot \mathbf{A}_{AS} \cdot \hat{S} + \hat{I}_B \cdot \mathbf{A}_{BS} \cdot \hat{S} \tag{19}$$

Through the use of the Wigner-Eckard theorem, Chao[31], Bullugiu[32, 33], then Scaringe et al.[34] in a more complete manner calculated the relations between the tensors of (19) and those of (2) + (3) and derived the general expressions:

$$\mathbf{g}_S = \frac{1+C}{2}\mathbf{g}_A + \frac{1-C}{2}\mathbf{g}_B \tag{20}$$

$$\mathbf{A}_{AS} = \frac{1+C}{2}\mathbf{A}_A \tag{21}$$

$$\mathbf{A}_{BS} = \frac{1-C}{2}\mathbf{A}_B \tag{22}$$

$$\mathbf{D}_S = \frac{C_1+C_2}{2}\mathbf{D}_A + \frac{C_1-C_2}{2}\mathbf{D}_B + \frac{1-C_1}{2}\mathbf{D}_{AB} \tag{23}$$

with

$$C = \frac{S_A(S_A+1) - S_B(S_B+1)}{S(S+1)} \tag{24}$$

$$C_1 = \frac{3[S_A(S_A+1) - S_B(S_B+1)]^2 + S(S+1)[3\,S(S+1) - 3 - 2\,S_A(S_A+1) - 2\,S_B(S_B+1)]}{(2\,S+3)(2\,S-1)S(S+1)} \tag{25}$$

$$C_2 = \frac{[4\,S(S+1) - 3][S_A(S_A+1) - S_B(S_B+1)]}{(2\,S+3)(2\,S-1)S(S+1)} \tag{26}$$

In the following sections, several applications of the relations (20)–(26) will be given. Our purpose now is to provide some general informations on the way to use these relations to interpret the magnetic and EPR data.

To obtain the theoretical expression for the magnetic susceptibility of A − B, we have to proceed as follows[35]:

(i) first determine the relative energies E(S) of the $^{2S+1}\Gamma$ states according to:

$$E(S) = -\frac{J}{2}S(S+1) \tag{27}$$

(ii) calculate the expressions for the $\mathbf{g}_S$ tensors from Eqs. (20) and (24). In fact, since most of the measurements are carried out on polycrystalline samples, we shall use average values g_S

(iii) take into account the zero field splitting within the ground state. Since the energy gaps between the pair states are large, the zero field splittings within the excited states will not play any significant role in the magnetic properties

(iv) finally calculate the susceptibility through the Van Vleck equation. Many examples will be given in Sect. 5 to 7.

In EPR, it is in principle possible to detect several pair states by varying the temperature, hence to determine for each of them the $\mathbf{g}_S$ and $\mathbf{D}_S$ (and eventually $\mathbf{A}_{AS}$ and $\mathbf{A}_{BS}$) tensors and to check the relations between local and molecular tensors. In fact, in many cases, even when the excited states are in principle EPR active and thermally populated, they are not detected. This is likely due to the fact that in the temperature range where an excited state is populated, the relaxation time is very short. Of course, when the total number of unpaired electrons $n_A + n_B$ is even, all the electronic degeneracies are generally removed in zero field and the system is then usually EPR silent. The Cu(II)Mn(II) pair discussed in Sect. 7.1 is a remarkable exception. When both n_A and n_B are odd, the EPR silence is in fact a good clue for the presence of interaction in A – B (see Sect. 5).

4 Symmetry of the Magnetic Orbitals and Nature of the Interaction

4.1 Comparison Cu(II)VO(II) and Cu(II)Cu(II)

A classical example which emphasizes the role of the symmetry of the magnetic orbitals is provided by the comparison of the two complexes $CuVO(fsa)_2en \cdot CH_3OH$ and $Cu_2(fsa)_2en \cdot CH_3OH$, noted [CuVO] and [CuCu] respectively. Their structures are shown in Fig. 1[36, 37]. For both complexes, the copper(II) ion occupying the inside site is in a square pyramid with a methanol molecule at the apex. In [CuVO], the vanadium(IV) of the vanadyl ion is also square pyramid whereas the copper(II) ion occupying the outside site of [CuCu] has planar geometry. For both [CuVO] and [CuCu], there is a pseudo mirror-plane containing the metal ions, perpendicular to the plane of the binucleating ligand. The unpaired electron of each Cu(II) ion is described by a magnetic orbital which is antisymmetric with respect to this mirror-plane and the unpaired electron for the VO(II) ion is described by a magnetic orbital which is symmetric with respect to the same mirror-plane. For both compounds, the interaction between the two local doublet states gives rise to a singlet and a triplet pair states. As schematized in Fig. 2, the two magnetic

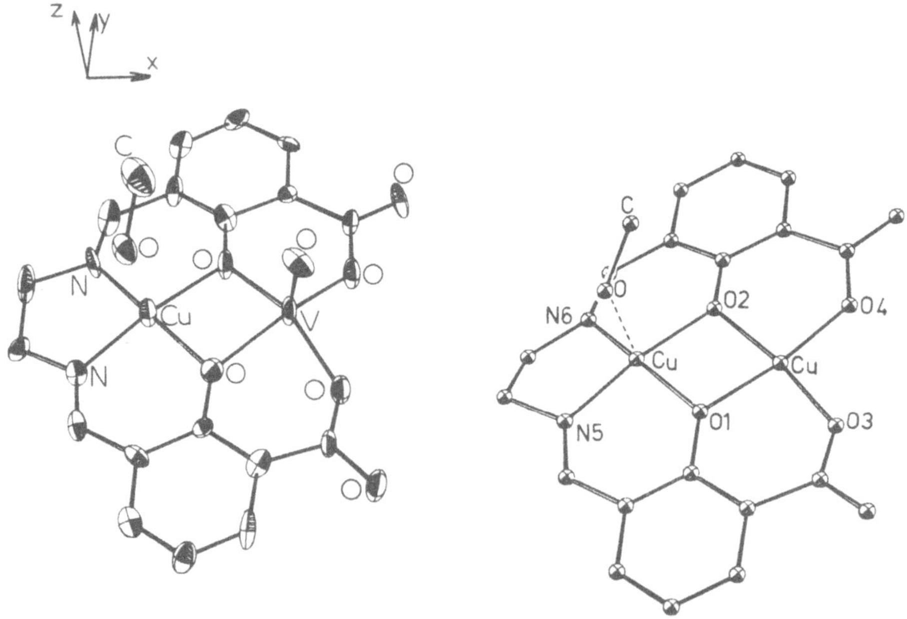

Fig. 1. Crystal structure of $CuVO(fsa)_2en \cdot CH_3OH$ and $Cu_2(fsa)_2en \cdot CH_3OH$ (from Ref. 36)

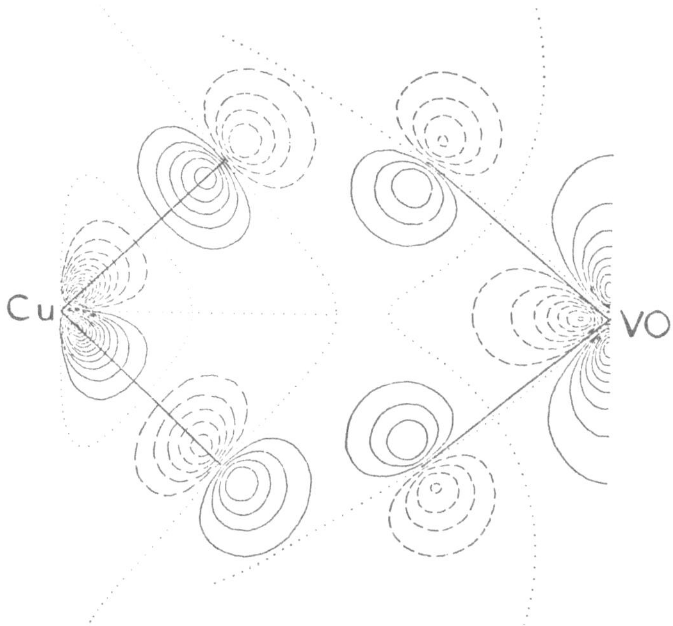

Fig. 2. Magnetic orbitals ϕ_{Cu} and ϕ_{VO} in the bridging zone of $CuVO(fsa)_2en \cdot CH_3OH$ (from Ref. 36)

orbitals in [CuVO] are strictly orthogonal; consequently the overlap is zero and the isotropic interaction is expected to be purely ferromagnetic with a triplet ground state stabilized by J with respect to the singlet excited state. J may be written as:

$$J = 2\,j = 2 \iint_{space} \frac{\varrho_{CuVO}(i)\varrho_{CuVO}(j)}{r_{ij}}\, d\tau(i)d\tau(j) \tag{28}$$

with:

$$\varrho_{CuVO}(i) = \phi_{Cu}(i)\phi_{VO}(i) \tag{29}$$

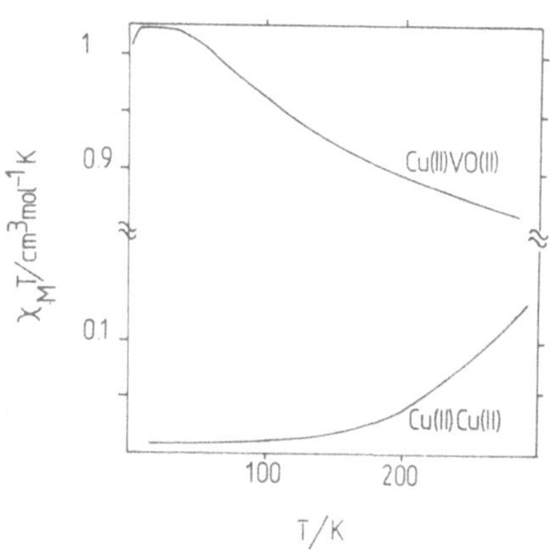

Fig. 3. $\chi_M T$ versus T plots for $CuVO(fsa)_2en \cdot CH_3OH$ and $Cu_2(fsa)_2en \cdot CH_3OH$ (from Ref. 36 and 37)

ϕ_{Cu} and ϕ_{VO} are the magnetic orbitals and ϱ_{CuVO} is defined as the overlap density between these magnetic orbitals[19]. In [CuCu], the two magnetic orbitals have the same symmetry and may overlap. We have seen that the antiferromagnetic contribution generally takes over the ferromagnetic one and the ground state is expected to be the singlet. The magnetic and EPR data confirm these predictions. The magnetic curves are compared in Fig. 3 in the form of $\chi_M T$ versus T plots, χ_M being the molar magnetic susceptibility and T the temperature. For [CuVO], $\chi_M T$ increases upon cooling down, which shows that the ground state has the highest spin multiplicity, then reaches a plateau corresponding to the temperature range where only the triplet ground state is thermally populated ($\chi_M T = 2\,N\beta^2 g^2/3\,k$). For [CuCu], $\chi_M T$ continuously decreases upon cooling down and reaches a value very close to zero, confirming that the ground state is now diamagnetic. Neglecting the zero field splitting within the triplet state, the same expression of $\chi_M T$ is valid for [CuVO] and [CuCu], namely:

$$\chi_M T = \frac{2\,N\beta^2 g^2}{k}\left[3 + \exp\left(-\frac{J}{kT}\right)\right]^{-1} \tag{30}$$

where N is Avogadro number, k the Boltzman constant and g the average value of the g-factor for the triplet state. From Eqs. (20) and (24), we have:

$$\mathbf{g} = (\mathbf{g}_{Cu} + \mathbf{g}_{VO})/2 \tag{31}$$

By fitting the magnetic data the singlet-triplet energy gap J is found equal to $+\,120\ \mathrm{cm}^{-1}$ in [CuVO] and $-\,650\ \mathrm{cm}^{-1}$ in [CuCu]. At 4.2 K, [CuVO] exhibits an X-band powder EPR spectrum that is typical of a triplet state with a zero field splitting characterized by an axial parameter $|D| = 3\,|D_{zz}|/2 = 0.24\ \mathrm{cm}^{-1}$ and a rhombic parameter $|E| = |D_{xx} - D_{yy}|/2 = 0.04\ \mathrm{cm}^{-1}$. The D_{uu}, u = x, y, z are the principal values of $\mathbf{D}_{CuVO}$. In contrast, below 77 K, [CuCu] is EPR silent.

4.2 Strict Orthogonality and Overlap Density

The orthogonality of the magnetic orbitals in [CuVO] explains why the triplet state is the lowest but does not provide any information on the magnitude of the phenomenon. Equation (28) suggests a topological approach of this problem.

Let A − B be a bimetallic compound in which the two magnetic orbitals ϕ_A and ϕ_B are orthogonal. The magnitude of the two-electron exchange integral j is governed by the zones of the space where the overlap density $\varrho_{AB}(i) = \phi_A(i)\phi_B(i)$ has a large absolute value. In such zones, the numerator of j, defined as in Eq. (28), is large and the denominator is small. Such a topological approach is illustrated in a schematic way in Fig. 4 which shows three situations (a), (b) and (c), all of which correspond to strict orthogonality of the magnetic orbitals. The zones of positive overlap density exactly compensate the zones of negative overlap density, so that the overlap integral $S = \int_{space}\varrho_{AB}(i)d\tau(i)$ is exactly zero. In (a), both the positive zone and the negative zone of ϱ_{AB} are weakly pronounced and the ferromagnetic interaction will be weak. (b) corresponds to an intermediate situation and (c) to a strong ferromagnetic interaction. Indeed, ϱ_{AB} exhibits a strongly positive zone, like a high mountain and a strongly negative zone,

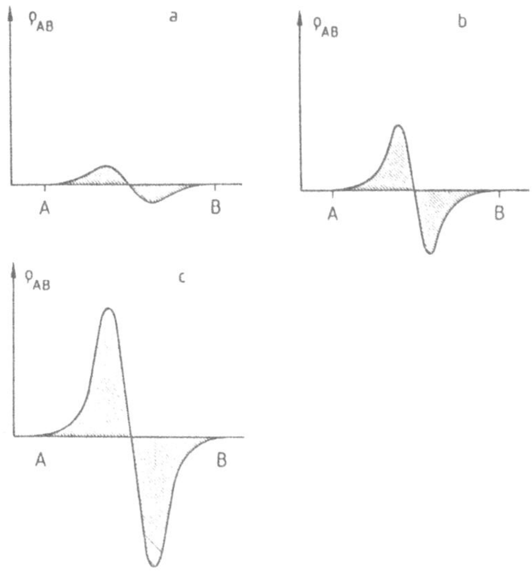

Fig. 4a–c. Schematic representation of the overlap density in three cases of strict orthogonality of the magnetic orbitals, leading to a weak **(a)**, intermediate **(b)** and strong **(c)** ferromagnetic interaction

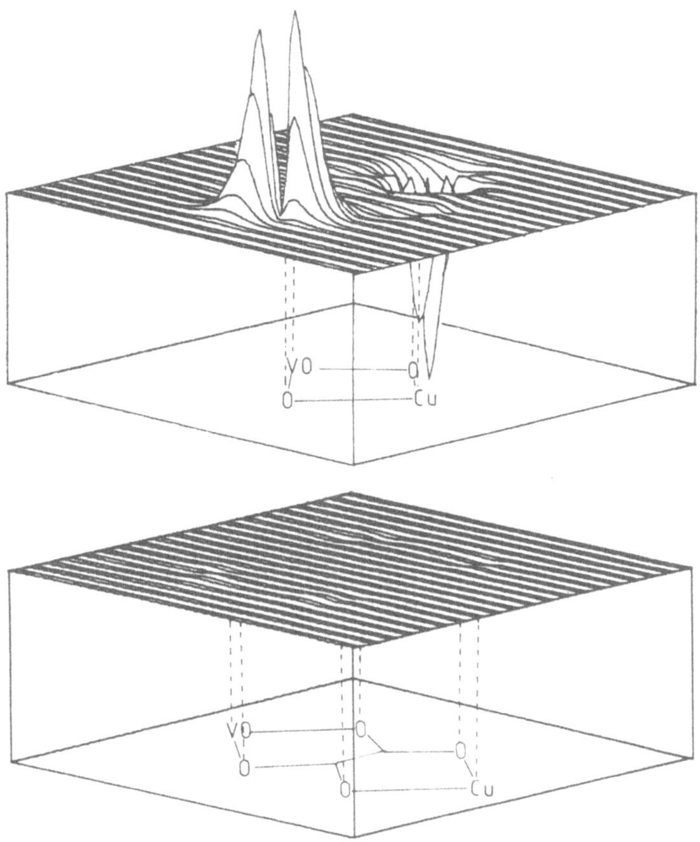

Fig. 5. Overlap density ϱ_{CuVO} in $CuVO(fsa)_2en \cdot CH_3OH$ *(top)* and $[(tmen)(H_2O)Cu(C_2O_4)VO(H_2O)(C_2O_4)]$ *(bottom)*

like a very deep hollow. For [CuVO], when plotting the overlap density from the magnetic orbitals of Fig. 2, one obtains the relief map of Fig. 5 with two strongly positive lobes around one of the oxygen bridging atoms and two strongly negative lobes around the other. This very favorable situation for achieving a strong ferromagnetic interaction is due to the fact that around each bridge the 2p oxygen orbital involved in ϕ_{Cu} has much the same orientation as the 2p oxygen orbital involved in ϕ_{VO}. Both point along the Cu–O direction.

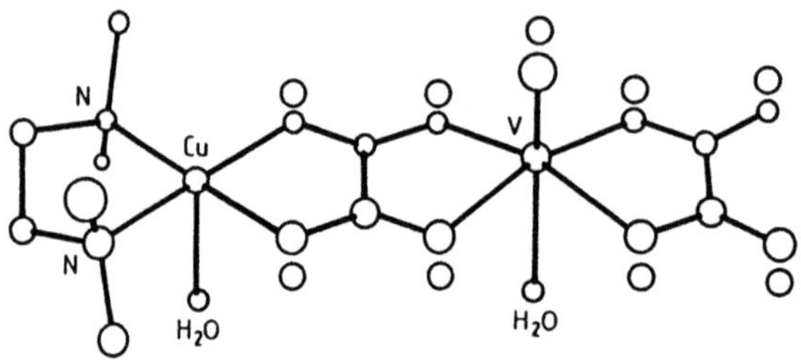

Fig. 6. Structure of $[(tmen)(H_2O)Cu(C_2O_4)VO(H_2O)(C_2O_4)]$ with tmen = N,N,N′,N′-tetramethylethylenediamine and C_2O_4 = oxalato (from Ref. 38)

This topological approach cannot be quantitative but seems to be very useful for comparing similar compounds. For instance, another Cu(II)VO(II) complex is represented in Fig. 6[38]. Again, Cu(II) and VO(II) are in square planar surroundings. Again, there is a mirror-plane perpendicular to the molecular plane and containing the metal ions. Again, the magnetic orbital around Cu(II) is antisymmetric and that around VO(II) is symmetric with regard to this mirror-plane. Again, the strict orthogonality of the magnetic orbital is achieved. But, when plotting the overlap density at the same scale as that used for [CuVO], we obtain the bottom map of Fig. 5. This map appears completely flat. The overlap density is spread over the whole extended bridging network and as a matter of fact, the ferromagnetic stabilization, if any, is smaller than 1 cm^{-1}.

4.3 Accidental Orthogonality

The strict or symmetry-imposed orthogonality is not the only type of orthogonality of the magnetic orbitals leading to a ferromagnetic interaction. The orthogonality may also be accidental[17, 39]. This situation, which occurs in a few heterobinuclear complexes (see Sect. 6.5), is easier to present in the case of planar bibridged copper(II) dimers like those schematized below. From two magnetic orbitals as shown in the left of Fig. 2, one can plot an overlap density map exhibiting two negative lobes along the x axis and two positive lobes along the y axis around each bridge. When the $\widehat{CuXCu}$ bridging angle is much larger than 90°, the overlap density along the x axis predominates. When $\widehat{CuXCu}$

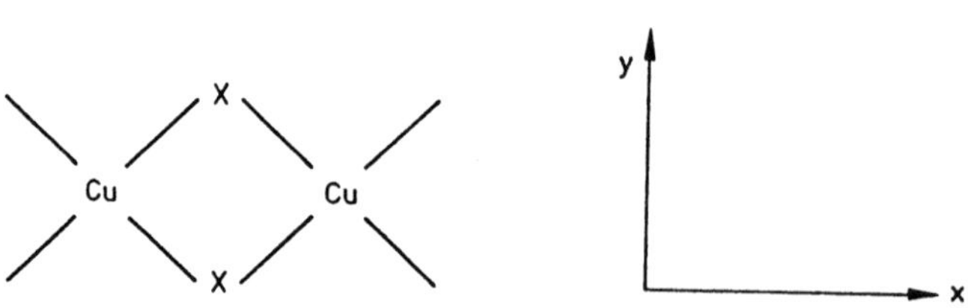

decreases, the balance between negative and positive lobes is less and less pronounced and for a value of $\widehat{CuXCu}$ close to 90° (but which cannot be predicted exactly a priori),the negative and positive lobes of the overlap density compensate around each bridge so that the overlap integral between the two magnetic orbitals is accidentally zero. Such a situation is represented in Fig. 7[30]. Several copper(II) and nickel(II) complexes and linear chains exhibiting a ferromagnetic interaction due to this accidental orthogonality have been reported[40–48].

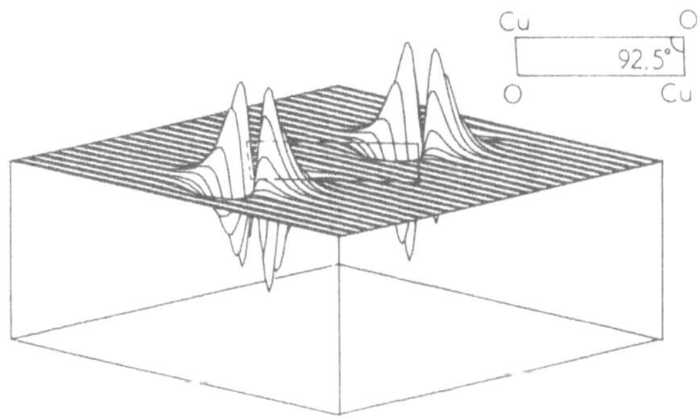

Fig. 7. Overlap density ϱ_{CuCu} for a bibridged copper(II) dimer in case of accidental orthogonality of the magnetic orbitals (from Ref. 30)

4.4 *Comparison Cu(II)Fe(III) and Cu(II)Cr(III)*

The concept of strict orthogonality can be extended to systems with more than two magnetic electrons. In this respect, the comparison of the two complexes $CuFe(fsa)_2en$-$Cl(H_2O) \cdot CH_3OH$ and $[CuCr(fsa)_2en(H_2O)_2]Cl$, noted [CuFe] and [CuCr] respectively, is particularly interesting from a pedagogical view point[49]. The structure of these complexes is shown in Fig. 8. Although not rigorously exact for [CuFe], we assume that the symmetry of the metal sites as well as the whole symmetry are close to C_{2v}. In [CuFe], there is just one unpaired electron around Cu(II), occupying a b_1 magnetic orbital. As for

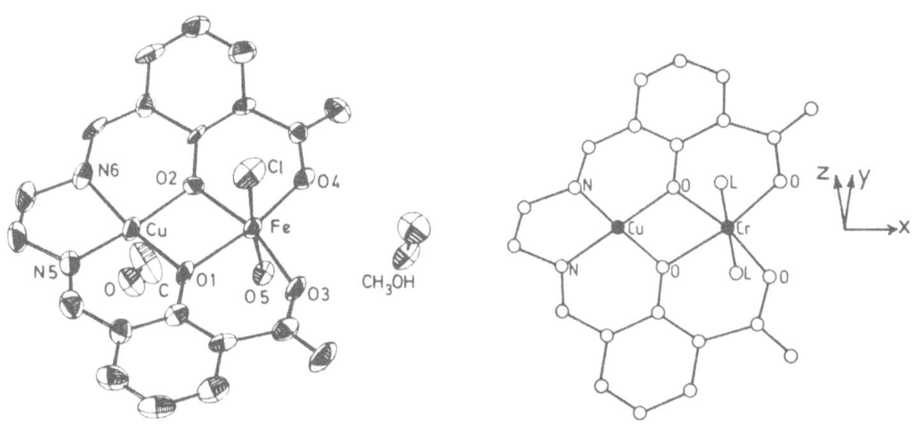

Fig. 8. Structures of $CuFe(fsa)_2enCl(H_2O) \cdot CH_3OH$ and $[CuCr(fsa)_2en(H_2O)_2]^+$ (from Ref. 49)

the Fe(III) ion, it is in a high-spin local state with five unpaired electrons occupying the five d-type magnetic orbitals and transforming as $a_1(x^2 - y^2$ and $z^2)$, $a_2(yz)$, $b_1(xy)$and $b_2(xz)$. One of these five magnetic orbitals, b_1, has the same symmetry as the only magnetic orbital around copper(II) and may overlap with it. This provides an antiferromagnetic contribution that is usually expected to be dominant. More generally, when one of the interaction centers of a polymetallic system has a high-spin d^5 configuration (Fe(III) or Mn(II)), the strict orthogonality cannot be achieved. At least one of the five d-type magnetic orbitals will overlap with a magnetic orbital centered on a nearest neighbor metal ion. In [CuFe], the isotropic interaction between the $^2B_1$ and $^6A_1$ local ground states leads to the $^5B_1$ and $^7B_1$ pair states, with the former expected to be the lowest. The quintet-septet energy gap is 3 J. From Eq. (6), J may be expressed as a sum of $J_{\mu\nu}$ contributions involving pairs of magnetic orbitals according to:

$$J = (J_{b_1a_1} + J_{b_1a'_1} + J_{b_1a_2} + J_{b_1b_1} + J_{b_1b_2})/5 \tag{32}$$

in which the only negative term $J_{b_1b_1}$ is dominant.

In [CuCr], the three unpaired electrons around Cr(III) in octahedral surroundings occupy the three t_{2g}-type magnetic orbitals transforming as $a_1(x^2 - y^2)$, $a_2(yz)$ and $b_2(xy)$. All three are orthogonal to the b_1 magnetic orbital around Cu(II). Therefore, in the expression of J:

$$J = (J_{b_1a_1} + J_{b_1a_2} + J_{b_1b_2})/3 \tag{33}$$

all the $J_{\mu\nu}$ contributions are purely ferromagnetic. The local ground states are $^2B_1$ and $^4B_1$, which leads to the $^2A_1$ and $^5A_1$ low lying pair states, the latter being now the lowest. The triplet-quintet energy gap is 2 J.

What is quite remarkable in this [CuFe] and [CuCr] comparison is that the two complexes are expected to have a same spin-quintet ground state. In the former case this ground state arises from an antiferromagnetic interaction owing to the overlap between two b_1 magnetic orbitals. In the latter case, it arises from a ferromagnetic interaction owing to the strict orthogonality of the four magnetic orbitals involved in the problem. The magnetic data schematized in Fig. 9 confirm these predictions. Upon cooling down,

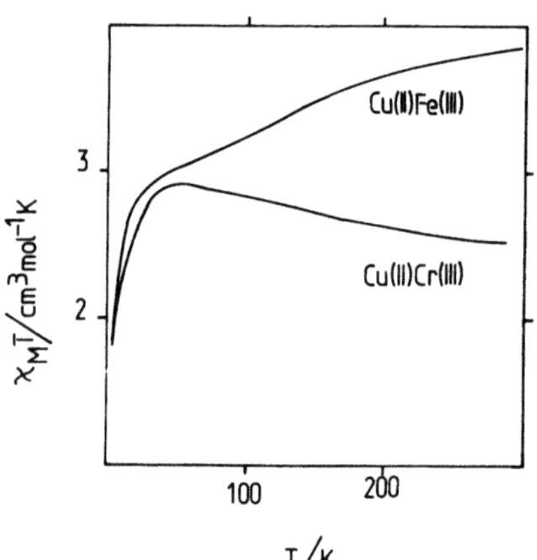

Fig. 9. $\chi_M T$ versus T plots for $CuFe(fsa)_2enCl(H_2O) \cdot CH_3OH$ and $[CuCr(fsa)_2en(H_2O)_2]Cl$ (from Ref. 49)

$\chi_M T$ decreases for [CuFe] and increases for [CuCr]. Below 50 K, the two curves are almost identical. This corresponds to the temperature range where, for both compounds, only the S = 2 ground state in thermally populated. The rather abrupt decrease of $\chi_M T$ in the very low temperature range is due to a strong zero field splitting of the ground states. As an example, we give some details on the calculation of the magnetic susceptibility for [CuFe]. $\mathbf{g}_2$ and $\mathbf{g}_3$ are related to $\mathbf{g}_{Cu}$ and $\mathbf{g}_{Fe}$ according to:

$$\mathbf{g}_2 = (7\,\mathbf{g}_{Fe} - \mathbf{g}_{Cu})/6$$
$$\mathbf{g}_3 = (5\,\mathbf{g}_{Fe} + \mathbf{g}_{Cu})/6 \tag{34}$$

The $\mathbf{D}_2$ tensor describing the splitting in zero field within the ground quintet state is related to $\mathbf{D}_{Fe}$ and $\mathbf{F}_{CuFe}$ as:

$$\mathbf{D}_2 = 4\,\mathbf{D}_{Fe}/3 - \mathbf{D}_{CuFe}/6 \tag{35}$$

Assuming that $\mathbf{D}_2$ is axial, we define an axial zero field splitting parameter $D = 3\,D_{2zz}/2$. The quintet state is then split as:

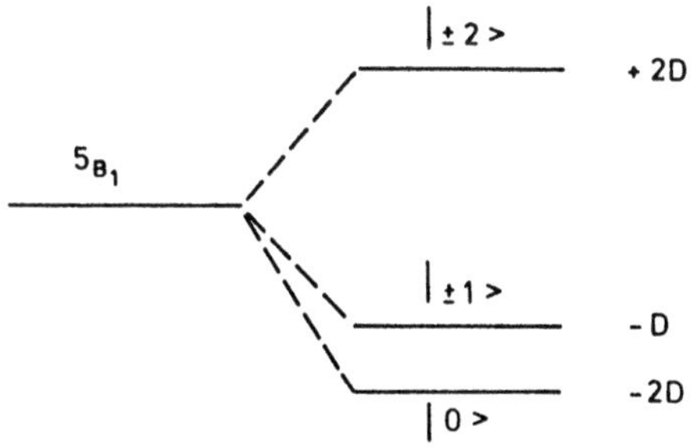

To the parallel and perpendicular susceptibilities within the quintet state, calculated using standard techniques[50)], one adds the susceptibility associated with the excited septet state and obtains the expressions:

$$\chi_{/\!/} = \frac{2\,N\beta^2}{kT}\{g_2^2\,[\exp(D/kT) + 4\exp(-2\,D/kT)] - 14\,g_3^2\exp(3\,J/kT)\}/$$
$$[\exp(2\,D/kT) + 2\exp(D/kT) + 2\exp(-2\,D/kT) + 7\exp(3\,J/kT)]$$

$$\chi_{\perp} = 2\,N\beta^2\left\{\frac{g_2^2}{3D}\,[9\exp(2\,D/kT) - 7\exp(D/kT) - 2\exp(-2\,D/kT)] + \frac{14\,g_3^2}{kT}\exp(3\,J/kT\right\}/$$
$$[\exp(2\,D/kT) + 2\exp(D/kT) + 2\exp(-2\,D/kT) + 7\exp(3\,J/kT)] \tag{36}$$

The spectra of the low lying states deduced from the magnetic data are shown in Fig. 10. Both [CuFe] and [CuCr] are EPR silent owing to the large zero-field splitting of the S = 2 state.

The interacting magnetic orbitals in [CuFe] (hence also in [CuCr]) are schematized in the first and third columns of Fig. 11 and the overlap densities between pairs of magnetic orbitals in the second column. $\varrho_{b_1a_1}$ is an overlap density of the same nature as that

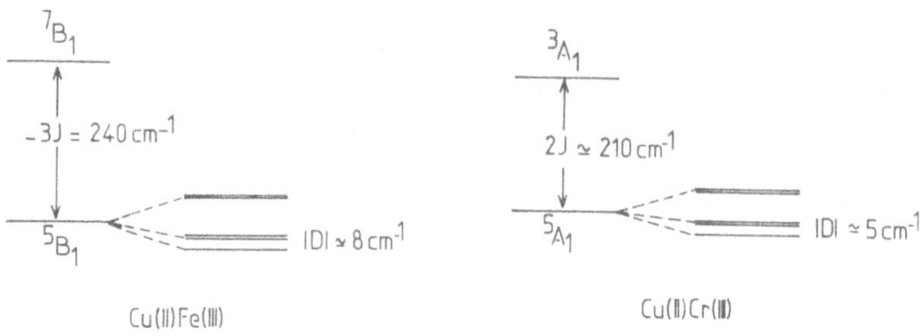

Fig. 10. Spectra of the low lying states for $CuFe(fsa)_2enCl(H_2O) \cdot CH_3OH$ and $[CuCr(fsa)_2en(H_2O)_2]Cl$ (from Ref. 49)

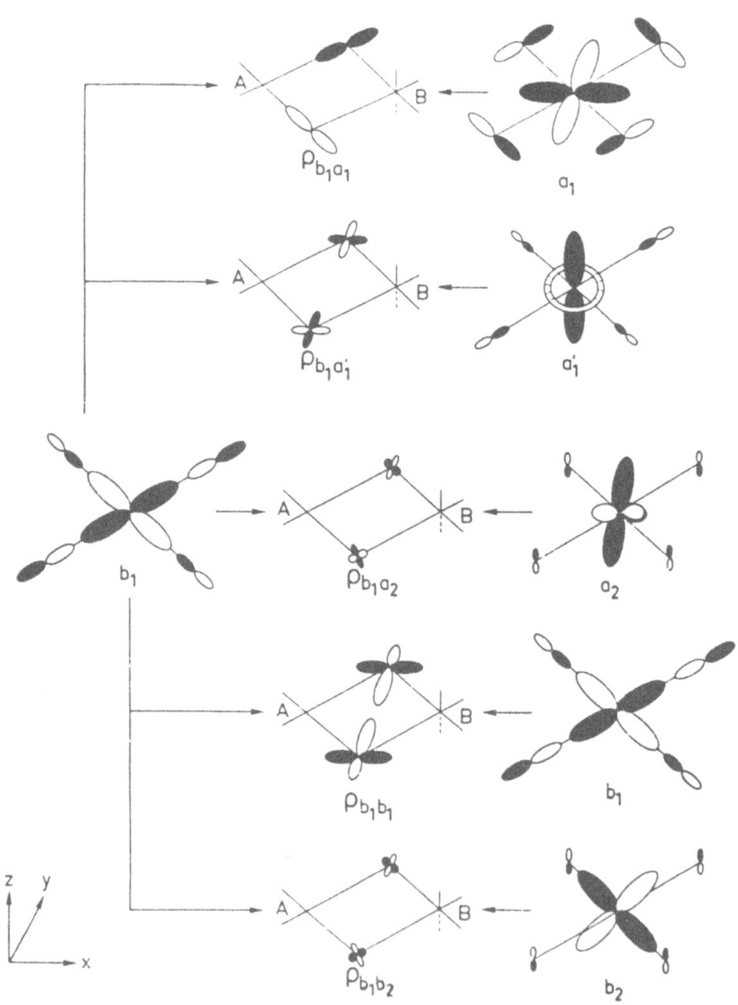

Fig. 11. Schematic representation of the magnetic orbitals and of the overlap densities between pairs of magnetic orbitals in [CuFe] and [CuCr] (from Ref. 49)

represented in Fig. 5 for [CuVO] with pronounced extrema leading to a strong ferromagnetic stabilization. $\varrho_{b_1a'_1}$ involves a z^2-type orbital weakly delocalized in the bridging network so that its extrema are less pronounced. $\varrho_{b_1a_2}$ and $\varrho_{b_1b_2}$ involve a_2 and b_2 orbitals which are very weakly delocalized in a π-manner on the bridging atoms. The xy plane is then a nodal plane and the ferromagnetic stabilization due to $\varrho_{b_1a_2}$ and $\varrho_{b_1b_2}$, if any, is certainly extremely weak. Finally, $\varrho_{b_1b_1}$ is as shown in Fig. 7. The overlap integral $S_{b_1b_1} = \int_{space}\varrho_{b1b1}(i)d\tau(i)$ is very sensitive to small structural changes. In the present case, the Cu O Fe bridging angles, equal to 99.4 and 100.7° respectively, are much larger than 90° and $S_{b_1b_1}$ is important enough to give a dominant antiferromagnetic interaction. With smaller bridging angles, the accidental orthogonality might in principle be obtained.

5 Cu(II)Fe(III) Pairs in Cytochrome Oxidase and Model Complexes

5.1 Oxidized Form of Cytochrome Oxidase

The study of the interaction between Cu(II) and high spin Fe(III) ions has attracted many researchers in the last few years. Their goal in general was not to understand the mechanism of the phenomenon in a thorough manner but to penetrate the *mystery* of the cytochrome oxidase, a respiratory enzyme reducing catalytically one mole of molecular oxygen into two moles of water[51)]. This enzyme contains two iron and two copper atoms per functioning unit. One iron is isolated in an heme unit; likewise one copper is also isolated. The second iron, again in an heme unit, interacts with the second copper to give in the oxidized form of the enzyme a Cu(II)Fe(III) bimetallic active site[52, 53)]. This Cu(II)Fe(III) pair is EPR silent[54, 55)] and the magnetic properties strongly suggest that up to 300 K, only a $S = 2$ pair state is thermally populated[56–58)]. Such a situation would arise from a very strong antiferromagnetic interaction between a Cu(II) and an high-spin Fe(III), as it occurs in the [CuFe] complex described in Sect. 4.4. The problem at hand is then to understand what kind of bridge is able to propagate such a strong antiferromagnetic interaction. So far, it has not been possible to recrystallize the enzyme and the hypotheses concerning the nature and the geometry of the bridge have evolved considerably over the last five years. Imidazolato from histidine[54, 56)] and thiolato groups[59–61)] among others have been proposed as a bridge. The latest EXAFS data suggest that the copper-iron separation is equal to 3.02 Å[62)]. Among the various hypotheses, this separation is particularly consistent with an oxygen bridging atom, belonging either to an oxo ion[63–64)], or to an hydroxo or a phenolato group.

Since the equatorial environment of the Fe(III) ion in the heme is imposed, the only available position is the axial one. In this case, the magnetic orbital around Fe(III) involved in the antiferromagnetic interaction is that of z^2 symmetry, which may be strongly delocalized toward the bridge owing to the antibonding overlap with the p_z ligand orbital. Two situations may be imagined, which lead to a strong overlap between this magnetic orbital and the only magnetic orbital around Cu(II): (i) the bridging atom belongs to the basal plane of the copper in a planar or an elongated tetragonal environment. The magnetic orbital around copper then points from the metal toward the four nearest neighbors of the basal plane, as in the [CuB] complexes studied above; (ii) the bridging atom belongs to the pseudo threefold axis of a copper in a trigonal bipyramidal environment. The magnetic orbital around copper is then of the z^2-type (z denoting here the axis of the bipyramid). These two situations with the orientation of the magnetic orbitals and the shape of the overlap densities are schematized in Fig. 12. None of these

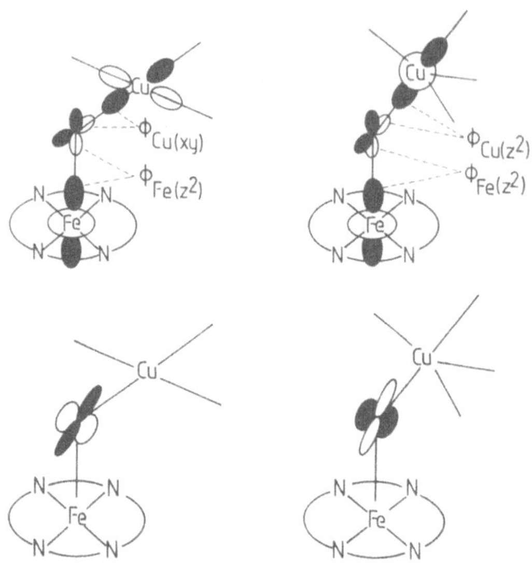

Fig. 12. Magnetic orbitals and overlap density in a Cu(II)XFe(III) network compatible with a strong antiferromagnetic interaction

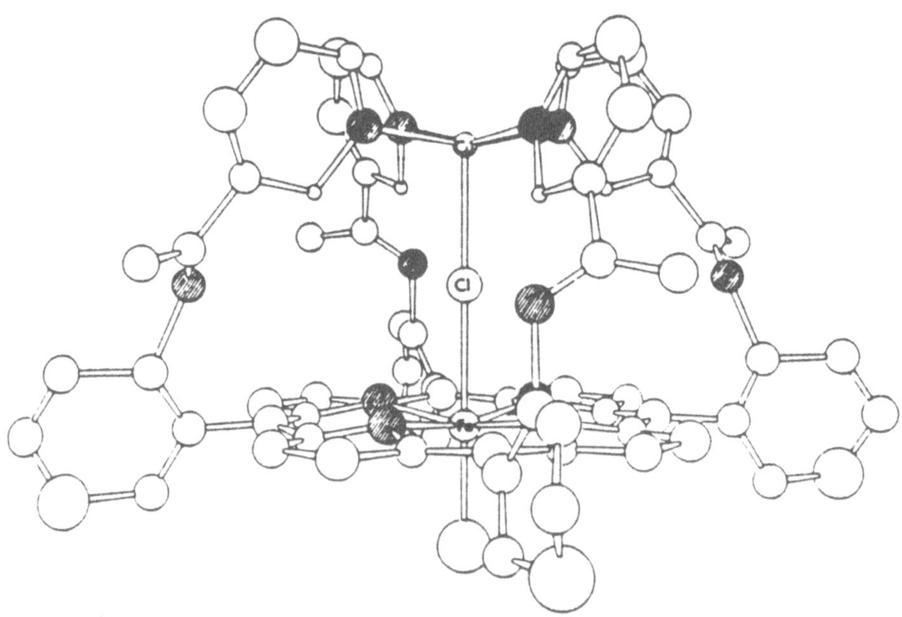

Fig. 13. Crystal structure of a Cu(II)Fe(III) complex presented as a model of cytochrome oxydase (from Ref. 66)

two structures have been designed so far, so that it is not surprising that the proposed model complexes[65–71)] do not exhibit the strong antiferromagnetic interaction apparently found in the enzyme. In fact, in a unique case, the structure of a model complex has been solved[65–66)]. This structure is depicted in Fig. 13. Copper is in a square pyramidal environment with a bridging chlorine atom occupying the apical position. The site symmetry of copper is very close to C_{4v}. In such a symmetry, there is no admixture of z^2-type orbital into the magnetic orbital centered on copper and transforming as b_1. It follows that there is no delocalization of the copper magnetic orbital on the chlorine bridge, hence a zero overlap density in any point of space. From the view point of the physics, this Cu(II)Fe(III) complex behaves as the superposition of the Cu(II) and Fe(III) ions in their individual surroundings. For instance, the EPR spectrum is essentially the superposition of the features characteristic of each of the metal centers. At this stage, a point maybe deserves to be pointed out. The technique of magnetic susceptibility measurement is extremely powerful when used properly. Not only the magnetic susceptibility itself but also the temperature of the sample (and not of somewhere else in the sample chamber) must be measured accurately since the quantity of interest to determine the nature of the low lying states is $\chi_M T$ (or μ_{eff} when using the old fashion notations). In the particular case of the Cu(II)Fe(III) bioinorganic models, the reported magnetic data are unfortunatly not always of sufficient accuracy to be interpreted unambiguously.

In addition to the actual model complexes where Fe(III) is in a porphyrin ring, several other Cu(II)Fe(III) complexes have been improperly presented as model complexes[64, 72, 73)]. For some of them, the claimed structure is debatable or even likely false, as it has been conclusively proved by Sinn in the case of the Cu(II)Fe(III) complexes with a 2,2′-bipyrimidine bridge[74)].

5.2 Cyanide Form of Cytochrome Oxidase

The nature of the low lying states in the oxidized form is not the only interesting problem dealing with cytochrome oxidase. By treating this form with cyanide, a new species is formed in which Fe(III) is now low spin. This cyanide form is again EPR silent. After some controversy, it seems now established, especially owing to a nice study in Magnetic Circular Dichroism[75–76)], that Cu(II) and low-spin Fe(III) are ferromagnetically coupled with a triplet ground state[77)]. The absence of EPR spectrum would be due to a strong zero field splitting within this triplet ground state.

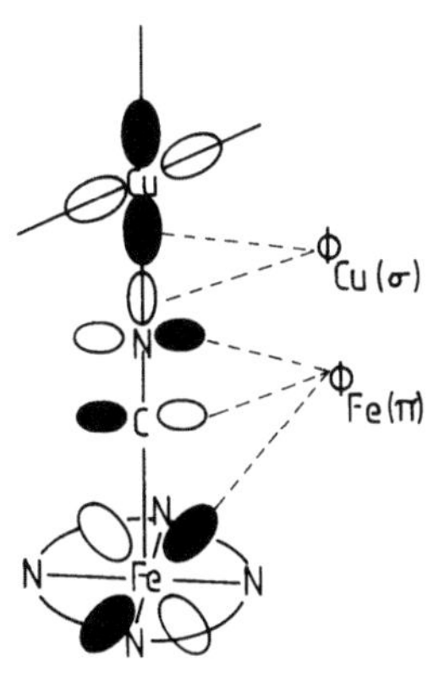

Fig. 14. Magnetic orbitals in a Cu(II)CNFe(III) network compatible with a ferromagnetic interaction

The origin of the ferromagnetic stabilization in this cyanide form can be rationalized by assuming a geometry as that of Fig. 14, where the nitrogen atom of CN^- belongs to the basal plane of the copper in an essentially planar environment, or alternatively to the axis of a trigonal bipyramid. The unpaired electron around Fe(III) is shared by the xy-, yz- and zx-type orbitals. The xy-type orbital is localized in the plane of the heme and cannot be involved in the Cu–Fe interaction. As for the other two orbitals, they have a π symmetry with respect to the FeCNCu axis. In contrast, the magnetic orbital around copper has a σ symmetry with respect to the same axis so that strict orthogonality is realized, as shown in Fig. 14. This orthogonality would be destroyed if the FeCNCu linkage was bent. Of course, if it is possible to account qualitatively for a ferromagnetic interaction with the geometry of Fig. 14, the expressions (6)–(10) are no longer appropriate to describe the phenomenon since one of the interacting ions is orbitally degenerate. This point will be discussed further in Sect. 8.3.

A model of this cyanide form of the cytochrome oxidase has been presented with a likely structure similar to that of Fig. 12, CN^- replacing Cl^- along the bridging linkage[78]. It is clear that such a geometry cannot give rise to a significant ferromagnetic interaction since again the magnetic orbital around copper is not delocalized toward the CN^- bridge.

6 Cu(II)Ni(II) Complexes

6.1 *Generalities*

To a large extent, the Cu(II)Ni(II) complexes with the local spins $S_{Cu} = 1/2$ and $S_{Ni} = 1$ represent ideal examples of heterobimetallic systems. Indeed, since one of the interacting ions is a spin doublet, the isotropic interaction leads to only two pair states, a doublet and a quartet separated by 3 J/2. When the interaction is antiferromagnetic the ground state is not split in zero field so that in the temperature range where only this ground doublet state is thermally populated, the magnetic susceptibility should follow a Curie law with $\chi_M T = N\beta^2 g_{1/2}^2/4\,k$. This allows an accurate determination of the average g-factor associated with the doublet state. The theoretical expression of the magnetic susceptibility, derived for the usual assumption that the isotropic interaction is much larger than the orther terms of the Hamiltonian (2), is easily derived as:

$$\chi_M T = \frac{N\beta^2}{4\,k}\,\frac{g_{1/2}^2 + 10\,g_{3/2}^2 \exp(3\,J/2\,kT)}{1 + 2\exp(3\,J/2\,kT)} \tag{37}$$

As for the EPR spectra, they are also expected to be relatively simple. Both the doublet state and the quartet state split in zero field into two kramers doublets are in principle EPR active. The relations (20)–(26) for the Cu(II)Ni(II) case give:

$$\mathbf{D}_{3/2} = (\mathbf{D}_{Ni} + \mathbf{D}_{CuNi})/3 \tag{38}$$

$$\begin{aligned} \mathbf{g}_{1/2} &= (4\,\mathbf{g}_{Ni} - \mathbf{g}_{Cu})/3 \\ \mathbf{g}_{3/2} &= (2\,\mathbf{g}_{Ni} + \mathbf{g}_{Cu})/3 \end{aligned} \tag{39}$$

Even the relations concerning the hyperfine structure are simple since only copper has isotopes, ^{63}Cu and ^{65}Cu, with a non zero nuclear spin. I_{Cu} is actually 3/2. These relations are:

$$\begin{aligned} A_{Cu\,1/2} &= -A_{Cu}/3 \\ A_{Cu\,3/2} &= A_{Cu}/3 \end{aligned} \tag{40}$$

6.2 $CuNi(fsa)_2en(H_2O)_2 \cdot H_2O$

The most studied Cu(II)Ni(II) complex is $CuNi(fsa)_2en(H_2O)_2 \cdot H_2O$, noted [CuNi][24, 79–82], of which the structure is shown in Fig. 15. Copper is in a strictly planar environment and nickel in a pseudo octahedral environment with two water molecules on each side of the molecular plane. The magnetic susceptibility data (Fig. 16) are characteristic of an antiferromagnetic interaction with a doublet ground state. Below 60 K, only this state is thermally populated and the magnetic susceptibility follows the Curie law with $\chi_M T = 0.52\ cm^3\ mol^{-1}$ K, corresponding to $g_{1/2} = 2.35$. Upon heating above 60 K, population of the quartet excited state begins and $\chi_M T$ progressively increases. The fitting of the magnetic data leads to a doublet-quartet energy gap of $3\,J/2 = -213\ cm^{-1}$, $g_{3/2}$ being equal to 2.23.

The X-band EPR properties of [CuNi] have been investigated on the pure material[81] and on magnetically dilute species[82]. In both cases, no signal associated with the quartet excited state has been detected, even above 60 K whereas the magnetic data confirm that

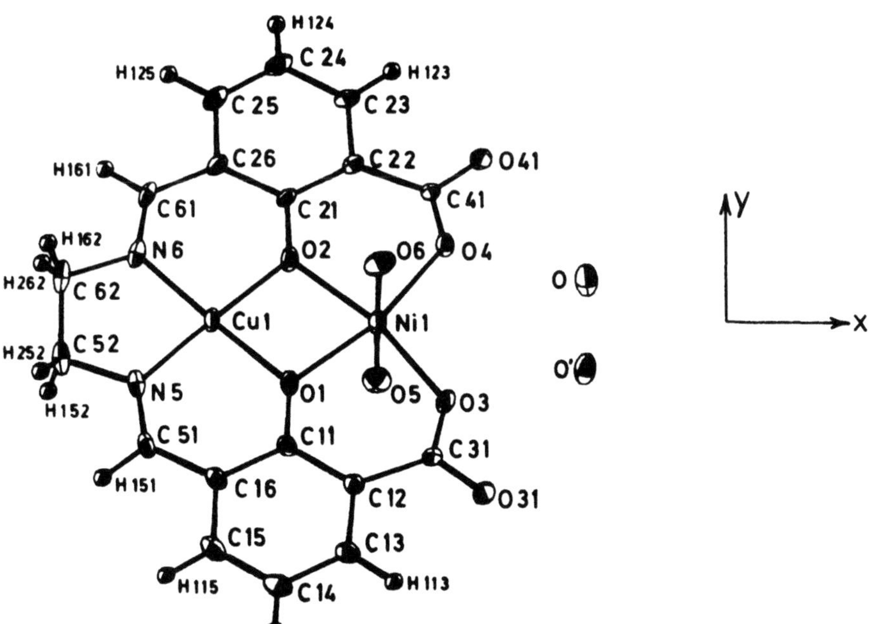

Fig. 15. Crystal structure of $CuNi(fsa)_2en(H_2O)_2 \cdot H_2O$ (from Ref. 81)

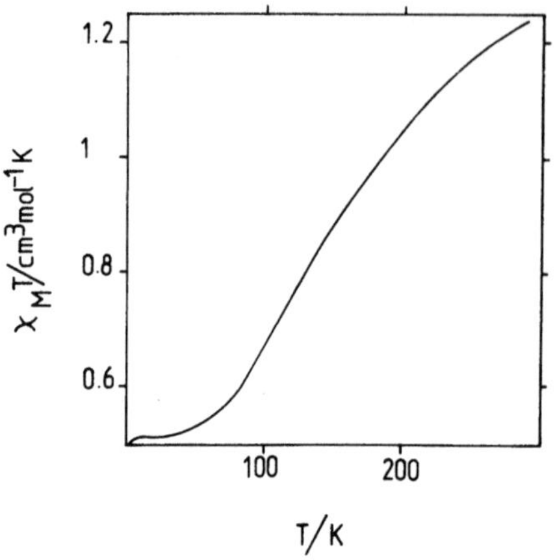

Fig. 16. $\chi_M T$ versus T plot for $CuNi(fsa)_2en(H_2O)_2 \cdot H_2O$ (from Ref. 81)

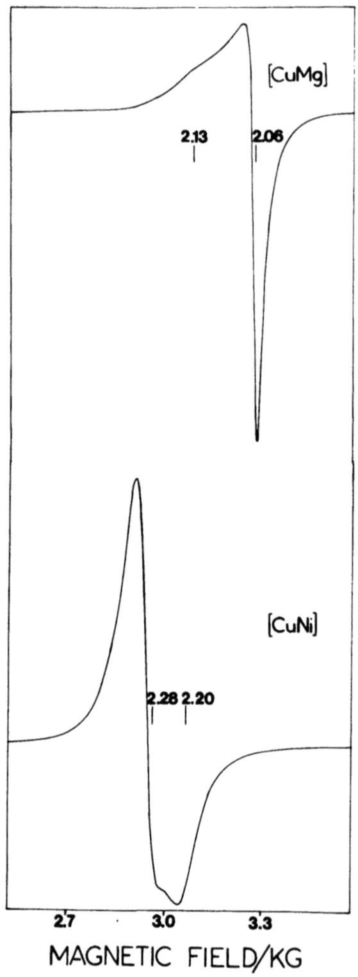

Fig. 17. Powder X-band EPR spectra of [CuMg] and [CuNi] (from Ref. 81)

this state is thermally populated. We already mentioned that this could be due to the fact that above 60 K, the relaxation time is too short. The EPR spectrum of the pure material is typical of a doublet state with $g_z = 2.23$ and $g_x \simeq g_y = 2.30$, without hyperfine structure, the axes being those of Fig. 15. It is interesting to point out that this doublet state cannot be confused with the local doublet state of copper(II). Indeed, for a Cu(II) ion in a planar environment, g_z is always larger than g_x and g_y. For instance, for [CuMg] of which the structure[83] is the same as that of [CuNi], one has $g_z = 2.13$ and $g_x \simeq g_y = 2.06$. On the contrary for [CuNi], g_z is the smallest principal value of $g_{1/2}$. (The powder EPR spectra of [CuMg] and [CuNi] are compared in Fig. 17.) This difference can be understood by examining the relation (39) between $\mathbf{g}_{1/2}$, $\mathbf{g}_{Ni}$ and $\mathbf{g}_{Cu}$. Since Ni(II) is in octahedral sorroundings, $\mathbf{g}_{Ni}$ may be assumed to be almost isotropic and the direction where $\mathbf{g}_{Cu}$ is maximum is that where $\mathbf{g}_{1/2}$ is minimum. Although [NiNi], where only the nickel(II) in the outside site is magnetic, does exist[81–84], the relation between $\mathbf{g}_{1/2}$, $\mathbf{g}_{Ni}$ and $\mathbf{g}_{Cu}$ cannot be experimentally checked. Indeed, [NiNi] is EPR silent so that g_{Ni} cannot be determined.

[CuNi] has been obtained in the form of dilute species by doping [CuCo] with about 1% of nickel(II). [CuCo] has a structure[85] very similar to that of [CuNi] and is EPR silent since it has an even number of unpaired electrons. In magnetically dilute [CuNi], the hyperfine structure is resolved as well as a small rhombicity of $\mathbf{g}_{1/2}$ in the plane of the binucleating ligand. The tensor $\mathbf{A}_{Cu\,1/2}$ has been determined as well as $\mathbf{A}_{Cu}$ by working with [CuCo] doped with Mg(II) and the coefficient 1/3 relating the two tensors has actually been found.

The main characteristic of the interaction in [CuNi] is the energy gap of $-213\ \mathrm{cm}^{-1}$ between ground doublet and excited quartet states. The antiferromagnetic nature of the interaction can be easily understood. The symmetry of the metal sites and of the complex as a whole is C_{2v}. The magnetic orbital around copper transforms as b_1 and those around nickel as a_1 and b_1, as depicted in Fig. 18. The CuONi bridging angle is equal to 99°, a value which is much above that corresponding to the accidental orthogonality of the b_1 magnetic orbitals. Therefore, the overlap between these orbitals gives rise to a predominant antiferromagnetic contribution $J_{b_1b_1}$ and J expressed as:

$$J = (J_{b_1a_1} + J_{b_1b_1})/2 \qquad (41)$$

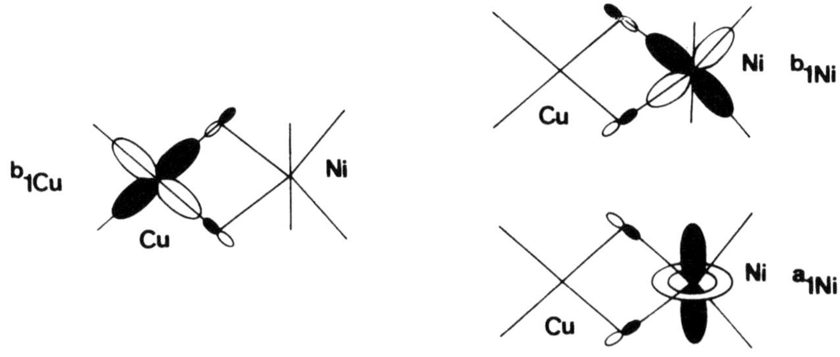

Fig. 18. Schematic representation of the magnetic orbitals centered on Cu(II) and Ni(II) in CuNi $(fsa)_2en(H_2O)_2 \cdot H_2O$ (from Ref. 81)

is negative. On can notice here, and this will be useful in Sect. 6.3, that there are two geometry changes able to reduce the magnitude of the antiferromagnetic interaction in a bibridged Cu(II)Ni(II) complex, either the decrease of the bridging angle, or the bending of the molecule along the axis joining the two bridging atoms. In both cases, the overlap between the b_1 magnetic orbitals lessens.

For another Cu(II)Ni(II) complex, both magnetic and EPR data have been reported[86], namely $CuNiLCl_2 \cdot 2\,H_2O$ where LH_2 is the binucleating Robson ligand[87]. Its likely structure is shown in Sect. 7.2. The doublet-quartet energy gap was found equal to $-309\ cm^{-1}$ and the EPR spectrum is actually characteristic of a doublet pair state with $g_{/\!/}$ (2.15) smaller than $g_{\perp}$ (2.28).

An antiferromagnetic interaction deduced from magnetic susceptibility measurements has also been detected in the adduct $Cu(Etsalen)Ni(hfa)_2$ where Etsalen is N,N′-bis(ethylsalicylaldiminato)ethylenediamine and hfa hexafluoroacetylacetonato[88]. The doublet-quartet energy gap was found equal to $-144\ cm^{-1}$. Unfortunately neither the crystal structure nor the EPR spectrum of this compound have been reported.

6.3 [Cu(salen)Ni(hfa)$_2$]$_2$; Complementary Between Magnetic and EPR Techniques

A compound quite similar to the last one discussed in Sect. 6.2 has been investigated in a thorough manner and has revealed quite a new aspect of the complementarity between the susceptibility and EPR techniques[89]. Its formula is $Cu(salen)Ni(hfa)_2$ and the structure of the heterobimetallic unit is shown in Fig. 19. As in the compound shown in Fig. 15, the copper is in planar and the nickel in octahedral surroundings, but the

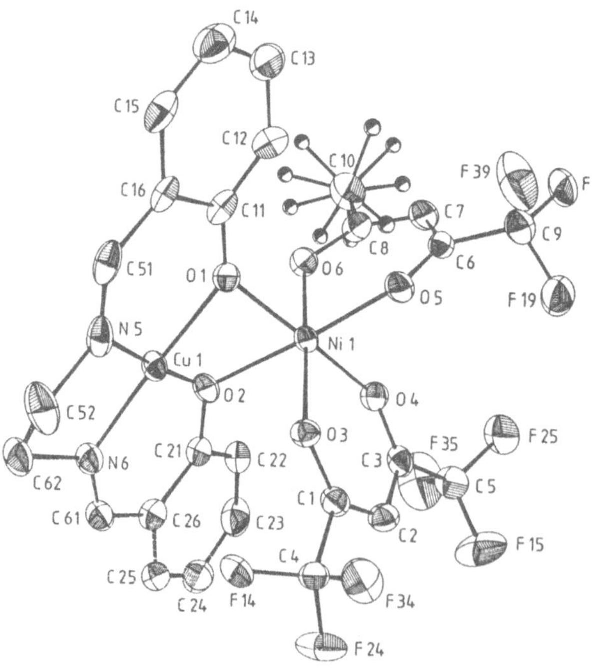

Fig. 19. Crystal structure of the $Cu(salen)Ni(hfa)_2$ unit (from Ref. 89)

CuO_2Ni network is bent along the axis joining the bridging atoms. The dihedral angle between the planes containing the bridging atoms is 141.4°. As pointed out in Sect. 6.2, this bending should lead to a decrease of the antiferromagnetic interaction and the magnetic data confirm this. The doublet-quartet energy gap is only $-35.4\ cm^{-1}$. Nevertheless, below 5 K and down to 1.3 K, the excited quartet state is totally depopulated and the magnetic susceptibility exhibits the Curie law expected for an isolated doublet state. Up to now, therefore, there is nothing of conceptually new. The situation changes when examining the powder EPR spectrum (Fig. 20). This spectrum, at 4.2 K, has definitively nothing to do with a simple doublet state and it exhibits all the characteristics of a triplet state split in zero field. As a matter of fact, it shows the $\Delta M_S = \pm 1$ allowed transitions as well as a $\Delta M_S = \pm 2$ forbidden transition at half field. This result, apparently inconsistent with the earlier discussion becomes clear when it is realized that, in the crystal lattice, the $Cu(salen)Ni(hfa)_2$ units are related two by two through a symmetry center to give rise to a remarkable bis heterobinuclear entity, as shown in Fig. 21. The interaction between two doublet pair states within a $[CuNi]_2$ entity leads to a singlet and a triplet with an energy separation that is too small to be deduced from the magnetic data, at least down to 1.3 K, although the triplet state is readily detected in EPR. In other words, the magnetic properties in the 1.3–300 K temperature range reflect the Cu(II)Ni(II) interaction within the heterobinuclear unit whereas the EPR *sees* the triplet state arising from the interaction between two doublet pair states within the bis heterobinuclear entity. A single crystal EPR study allowed the determination of the principal values of the **g** and **D** tensors associated with this triplet state and their principal orientations with regard to the $[CuNi]_2$ entity. The results confirm that the triplet state arises from an interaction between two doublet pair states and not from an interaction between the two local states of the symmetry related Cu(II) ions.

A last point concerning the compound $Cu(salen)Ni(hfa)_2$ deserves to be discussed, if not solved, namely why is the zero field splitting within the triplet state rather large with $|D| = 3\,|D_{zz}|/2 = 0.125\ cm^{-1}$ and $|E| = (|D_{xx} - D_{yy}|)/2 = 0.019\ cm^{-1}$ whereas the interaction between two symmetry related CuNi units is very small? In the present case, the zero field splitting may have three origins, the third being specific to this kind of interaction between two symmetry related Cu(II)Ni(II) pairs: (i) the magnetic dipolar

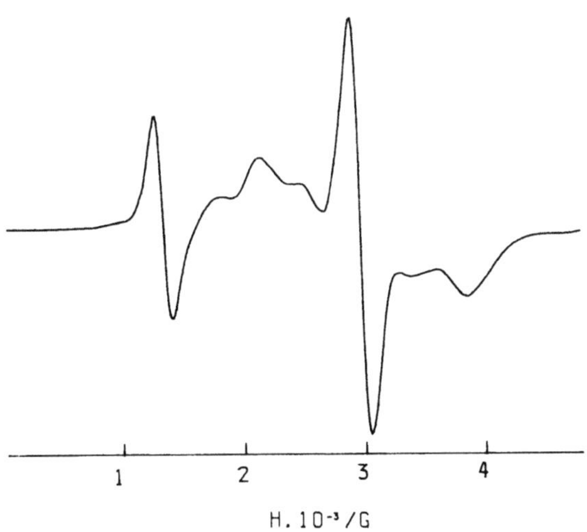

Fig. 20. Powder X-band EPR spectrum of $[Cu(salen)Ni(hfa)_2]_2$ (from Ref. 89)

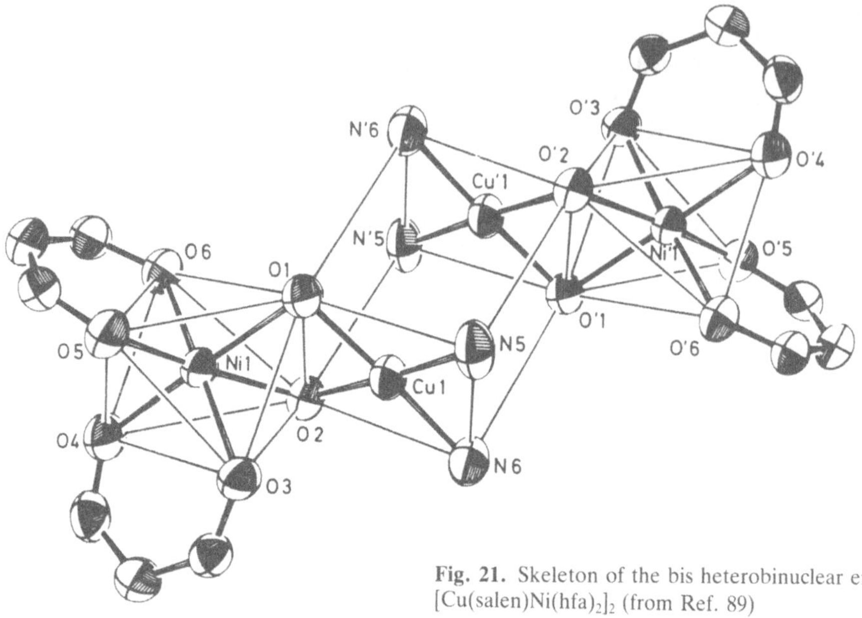

Fig. 21. Skeleton of the bis heterobinuclear entity $[Cu(salen)Ni(hfa)_2]_2$ (from Ref. 89)

interaction. By using the approximation of point magnetic dipoles, it has been shown that this contribution accounts for less than a tenth of the total phenomenon; (ii) the anisotropic interaction. We have seen that this contribution is proportional to the interaction between the ground state of one Cu(II)Ni(II) pair and the excited states of the other. Maybe, the excited pair state involving a z^2-type copper orbital (z being here the axis perpendicular to the copper basal plane) could interact more strongly with the ground pair state of the other CuNi unit than the two ground pair states do; (iii) the doublet-quartet mixing within a CuNi unit[90]. This new contribution comes from the circumstance that the ground state of a CuNi unit is not a rigorously pure spin doublet state. It contains a weak contribution from the spin quartet state located at 35.4 cm^{-1} above, owing to the coupling of the two low lying pair states through the local anisotropy of nickel(II). This mixing may contribute to the zero field splitting within the triplet state of a $[CuNi]_2$ entity. This new contribution varies as D_{Ni}/J on the one hand and as the singlet-triplet energy gap within $[CuNi]_2$ on the other hand. It has been demonstrated that with a

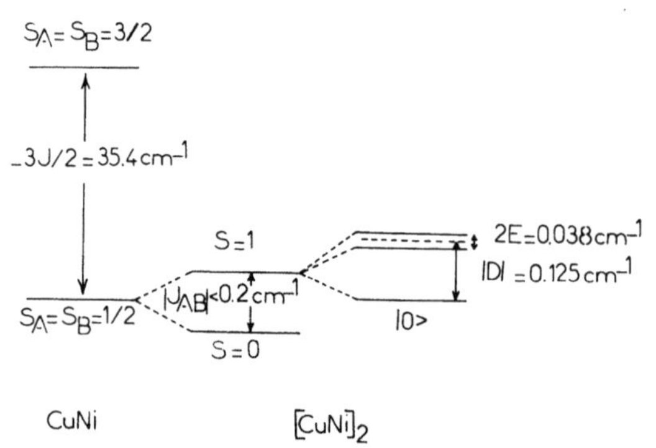

Fig. 22. Spectrum of the low lying states in $[Cu(salen)Ni(hfa)_2]_2$ (from Ref. 89)

singlet-triplet splitting of 0.2 cm^{-1}, which could not be detected from the magnetic data down to 1.3 K, a local anisotropy around nickel(II) characterized by an axial parameter $D_{Ni} = 5$ cm^{-1} would lead to the experimentally observed zero field splitting. Such a D_{Ni} value for the local anisotropy of nickel(II) is not at all exceptional[50)].

From magnetic and EPR data, the spectrum of the low lying states in [Cu(salen)-Ni(hfa)$_2$]$_2$ may be schematized as shown in Fig. 22.

6.4 A Cu(II)Ni(II) Complex with a Ferromagnetic Interaction

There is a unique example of Cu(II)Ni(II) complex in which the isotropic interaction has been reported to be ferromagnetic with a quartet ground state[91)], namely Ni(cth)$CuCl_4$ with cth = d,l-5,7,7,12,14,14-hexamethyl-1,4,8,11-tetraazacyclotetradecane. The crystal structure of this compound is not known but that of the isomorphous compound Ni(cth)$ZnCl_4$ has been solved. This structure is represented in Fig. 23 and the magnetic properties are shown in Fig. 24. Upon cooling, $\chi_M T$ increases, then reaches a plateau with $\chi_M T = 1.98$ cm^3 mol^{-1} K, a value consistent with an isolated quartet state ($\chi_M T = 5 N\beta^2 g_{3/2}^2/4k$), $g_{3/2}$ being then equal to 2.05. The doublet-quartet splitting has been estimated as 125 ± 25 cm^{-1}.

It is worth emphasizing here that the determination of J from the magnetic data is more accurate for an antiferromagnetic interaction ($J < 0$) than for a ferromagnetic

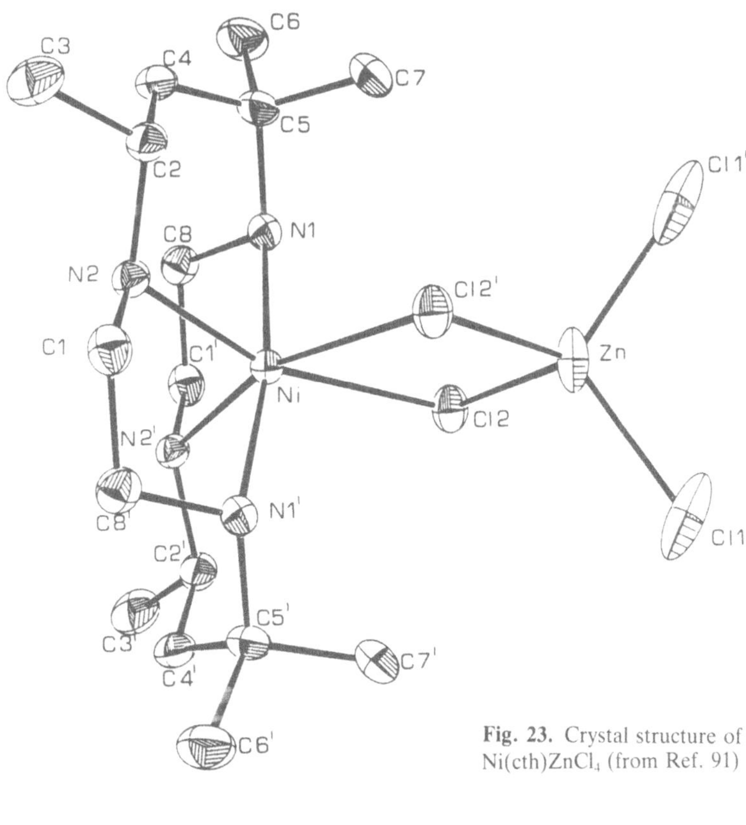

Fig. 23. Crystal structure of Ni(cth)$ZnCl_4$ (from Ref. 91)

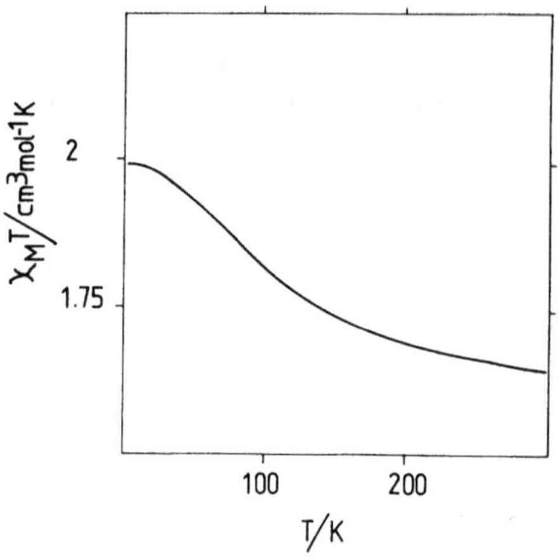

Fig. 24. $\chi_M T$ versus T plot for Ni(cth)$CuCl_4$ (from Ref. 91)

interaction ($J > 0$). Let us justify this important point from an experimental view point in the case of a Cu(II)Ni(II) pair. For sake of simplicity, we assume $g_{Cu} = g_{Ni} = g$. At very high temperature, when kT is much larger than $|J|$, the limit for $\chi_M T$ is:

$$(\chi_M T)_{HT} = \frac{N\beta^2 g^2}{3\,k} \times \frac{11}{4} \quad (42)$$

For $J < 0$, the low temperature limit corresponding to an isolated doublet state is:

$$(\chi_M T)_{LT} = \frac{N\beta^2 g^2}{3\,k} \times \frac{3}{4} \quad (43)$$

so that we have:

$$(\chi_M T)_{HT}/(\chi_M T)_{LT} = 11/3 \quad (44)$$

In other words, upon cooling, the $\chi_M T$ value decreases by 72.7%. For $J > 0$, the low temperature limit corresponding to an isolated quartet state is:

$$(\chi_M T)_{LT} = \frac{N\beta^2 g^2}{3\,k} \times \frac{15}{4} \quad (45)$$

and

$$(\chi_M T)_{HT}/(\chi_M T)_{LT} = 15/11 \quad (46)$$

hence, upon cooling, the $\chi_M T$ value only increases by 36.4%. This result is general. For any pair, the dependence of $\chi_M T$ on T plot is larger for $J < 0$ than for $J > 0$.

Surprisingly, Ni(cth)$CuCl_4$ does not exhibit any EPR signal, which could be due either to relaxation effects, or alternatively, the lowest Kramers doublet arising from the splitting in zero field of the ground quartet state is $|\pm 3/2\rangle$. This Kramers doublet may be not detectable in EPR owing to its anisotropy.

The most interesting problem related to this Cu(II)Ni(II) complex is the origin of the ferromagnetic interaction. Does the stabilization of the quartet state arise from a strict or an accidental orthogonality of the magnetic orbitals? To answer, let us consider the reference frame depicted in Fig. 25. The binuclear species has a C_2 symmetry. The two magnetic orbitals around Ni(II) are of the z^2- and xy-type, transforming as a and b irreducible representations respectively. Around copper(II) in a compressed tetrahedral environment, the magnetic orbital mixes xy- and yz-type orbitals transforming as b. It follows that J may be written as:

$$J = (J_{ba} + J_{bb})/2 \tag{47}$$

J_{ba} involving orthogonal magnetic orbitals is likely a small ferromagnetic contribution since the a magnetic orbital around Ni(II) is only weakly delocalized toward the chlorine bridging atoms. As for J_{bb}, it involves magnetic orbitals of the same symmetry and may yield ferro- or antiferromagnetic contribution, depending on the details of the structure. It is probable that in the present case, the two b magnetic orbitals are quasi accidentally orthogonal, yielding a ferromagnetic interaction.

6.5 *Magnetically Dilute Cu(II)Ni(II) Pairs*

In addition to the studies mentioned in Sect. 6.2–6.4 dealing with actual heterobimetallic complexes, several studies have been devoted to dilute species in host lattices. These species are generally easier to obtain than the pure Cu(II)Ni(II) compounds. Bibridged copper(II) binuclear complexes are often good host lattices. Indeed, most of them are strongly antiferromagnetically coupled so that they are EPR silent over a large temperature range. To our knowledge, Kokoszka has been the first to detect an heteropair by studying the EPR spectrum of nickel(II)-doped bis[μ-pyridine N-oxide dichloroaquocopper(II)][92–93]. Nickel(II) binuclear complexes that are EPR silent owing to the local anisotropy have also been used as host lattices.

All the authors working on dilute species obviously focused on the EPR properties so that if, in general, they could determine the nature of the isotropic interaction, they were

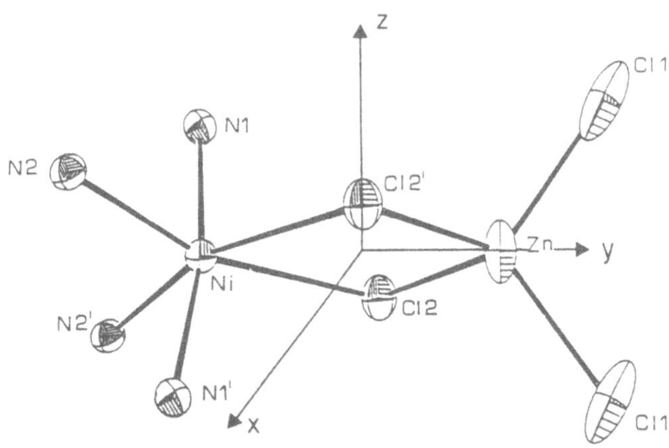

Fig. 25. Reference frame for $Ni(cth)CuCl_4$ (from Ref. 91)

not able to specify its magnitude. In principle, this magnitude might be deduced from the temperature dependence of the EPR signal though in fact, this method is usually quite unreliable. This is the case, for instance, for $CuNi(fsa)_2en(H_2O)_2 \cdot H_2O$ (see Sect. 3.2), both as a pure material and dilute species in [CuCo]. The product of the intensity of the signal associated with the doublet ground state with temperature is constant as expected in the 4.2–60 K temperature range when only this ground state is populated but above 60 K an inhomogeneous broadening of the signal occurs, which leads to a large uncertainty in the intensity measurements.

The main goal of the EPR studies on dilute Cu(II)Ni(II) species was to check the validity of the relations (38)–(40) between pair and local tensors[33, 82, 94–97]. If the relation (39) dealing with the **g** tensors is difficult to verify since $\mathbf{g}_{Ni}$ cannot be determined directly, that dealing with the hyperfine structure has been found to be valid in all the cases where the isotropic interaction is expected to be large. Let us give an example of this in addition to that reported in Sect. 6.2. The structure of $M_2(trik)(py)_4$ (where trik and py stand for 1,5-diphenyl-1,3,5-pentanetrianato and pyridine respectively) is:

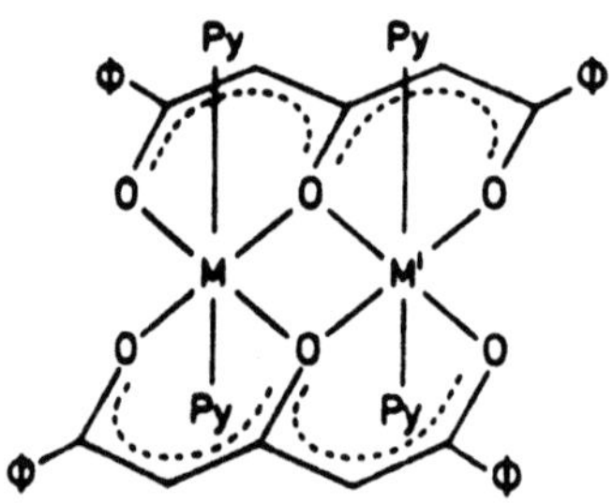

In the copper(II)-doped $Ni_2(trik)(py)_4$ compound with Cu(II)Ni(II) pairs[96], $A_{/\!/}$ was found equal to 45×10^{-4} cm^{-1} and in copper(II)-doped $Zn_2(trik)(py)_4$ compound with isolated Cu(II) ions, $A_{/\!/}$ is 136×10^{-4} cm^{-1}. In one case at least, however, the ratio of 1/3 between $A_{Cu\,1/2}$ and A_{Cu} has not been found[98], namely in nickel(II) (or zinc(II)) doped tetra-μ-benzoato bis[quinoline copper(II)]. The structure of the copper(II) binuclear complex is very similar to that of the famous copper(II) acetate. In the Cu(II)Ni(II) pair, $A_{/\!/}$ is 94×10^{-4} cm^{-1} and in the Cu(II)Zn(II) species $A_{/\!/}$ is 138×10^{-4} cm^{-1}. This result has been attributed by Bencini and Gatteschi[90] to the fact that in the Cu(II)Ni(II) pair, the isotropic interaction is not very large with regard to the local anisotropy of nickel(II). The doublet-quartet mixing through this local anisotropy may be relatively important and relation (40) may be no longer valid. Alternatively, the deviation of $\mathbf{A}_{Cu\,1/2}$ (or $\mathbf{A}_{Cu\,3/2}$) from $\mathbf{A}_{Cu}/3$ may be used to estimate the ratio D_{Ni}/J. In the present case, the observed hyperfine structure parameters are consistent with D_{Ni}/J close to the unity. Even when D_{Ni}/J is equal to 0.1, a deviation with regard to the ratio 1/3 can be experimentally detected. One will notice that this doublet-quartet coupling through D_{Ni} has also been postulated as one of the possible origins of the zero field splitting within the triplet state of $[Cu(salen)Ni(hfa)_2]_2$ (see Sect. 6.3).

7 Other Selected Examples of Heteropairs Without Orbital Degeneracy

In this section, we selected a few additional examples of heterobimetallic complexes in which the interacting ions have no first order angular momentum. This choice, as well as the organization of this article as a whole, reflects our subjectivity and does not mean in any way that no other interesting study along this line has been published.

7.1 A Dilute Cu(II)Mn(II) Pair

The first example deals with a Cu(II)Mn(II) pair dilute in bis[μ-pyridine N-oxide dichloroaquocopper(II)], studied by McPherson et al.[99–100]. The structure of the host lattice is shown in Fig. 26. It consists of two distorded square pyramids sharing a basal edge such that the dimer is centrosymmetric. This copper(II) dimer is strongly antiferromagnetically coupled with a singlet-triplet energy gap of $-880\ cm^{-1}$. The Mn(II)-doped species exhibits beautifully resolved EPR spectra (Fig. 27), with both fine and hyperfine structures. The spectrum consists of the four groups of lines expected for a $S = 2$ pair state with a small zero field splitting. To this fine structure is superimposed hyperfine structure due to the interaction with both the manganese and copper nuclei, the nuclear spins being 5/2 for ^{55}Mn and 3/2 for ^{63}Cu and ^{65}Cu. More precisely, each fine structure is split into six lines by the interaction with the Mn nucleus and each of these lines then split into four by the interaction with the Cu nucleus to produce the 24-line pattern seen in Fig. 27.

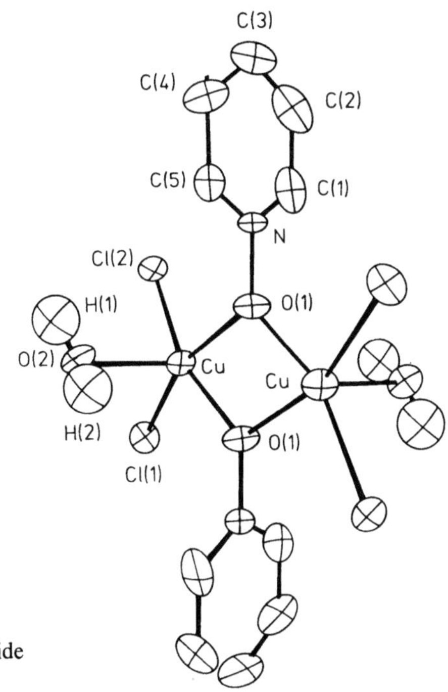

Fig. 26. Crystal structure of bis[μ-pyridine N-oxide dichloroaquocopper(II)] (from Ref. 100)

The relations (20)–(26) applied to the quintet state of the Cu(II)Mn(II) pair give:

$$\mathbf{g}_2 = (7\,\mathbf{g}_{Mn} - \mathbf{g}_{Cu})/6 \quad (48)$$

$$\mathbf{D}_2 = 4\,\mathbf{D}_{Mn}/3 - \mathbf{D}_{CuMn}/6 \quad (49)$$

$$\mathbf{A}_{Mn2} = 7\,\mathbf{A}_{Mn}/6 \quad (50)$$

$$\mathbf{A}_{Cu2} = -\,\mathbf{A}_{Cu}/6 \quad (51)$$

$\mathbf{g}_{Mn}$ is certainly almost isotropic with principal values very close to the free-electron value. Therefore, the principal values of $\mathbf{g}_2$ are expected to be smaller than 2.00 since those of $\mathbf{g}_{Cu}$ are all larger than 2.00. $\mathbf{g}_{Cu}$ is actually known from the EPR spectrum of the Zn(II)-doped species and the relation (48) is very nicely confirmed. The experimental principal values of $\mathbf{g}_2$ are $g_x = 1.991$, $g_y = 1.986$ and $g_z = 1.952$ and the calculated ones from $\mathbf{g}_{Cu}$ and $\mathbf{g}_{Mn}$ assumed to be isotropic with principal values equal to 2.002 are $g_x = 1.993$, $g_y = 1.988$ and $g_z = 1.951$. As for the hyperfine structure, relation (51) describing the interaction with the Cu nucleus is perfectly confirmed; the hyperfine structure parameter along the direction perpendicular to the bridging network is $139 \times 10^{-4}\ cm^{-1}$ in the Zn(II)-doped species and $23 \times 10^{-4}\ cm^{-1}$ in the Mn(II)-doped species. Since $\mathbf{A}_{Mn}$ is not known, relation (50) cannot be checked as rigorously as relation (51); nevertheless values of about $80 \times 10^{-4}\ cm^{-1}$ are reported for the hyperfine structure parameter of a Mn(II) ion surrounded by chloride ions and oxygen-containing ligands[101]. In the Mn(II)-doped species, a value of $91 \times 10^{-4}\ cm^{-1}$ is found, which is actually close to 7/6 the value for the non coupled Mn(II) ion. The quintet-septet energy gap cannot be determined, but whatever it may be, these two pair states do not couple significantly through the local anisotropy around Mn(II) since D_{Mn} is expected to be of the order of $10^{-1}\ cm^{-1}$ [102]. It is, therefore, not surprising that relations (48)–(51) are so precisely verified, in contrast with the example discussed in Sect. 6.5. Finally the zero field splitting within the quintet state is characterized by the parameters $|D| = 0.051\ cm^{-1}$ and $|E| = 0.013\ cm^{-1}$, the orientation of $\mathbf{D}_2$ being quite different from that of $\mathbf{g}_2$. The anisotropy of $\mathbf{g}_2$ is that of $\mathbf{g}_{Cu}$ with the z axis perpendicular to the basal plane and the x and y axes roughly along the directions of the bonds in this basal plane while for $\mathbf{D}_2$, the z axis lies within 5° of the plane of the bridging network and the other two axes are not directed along any particular structural features.

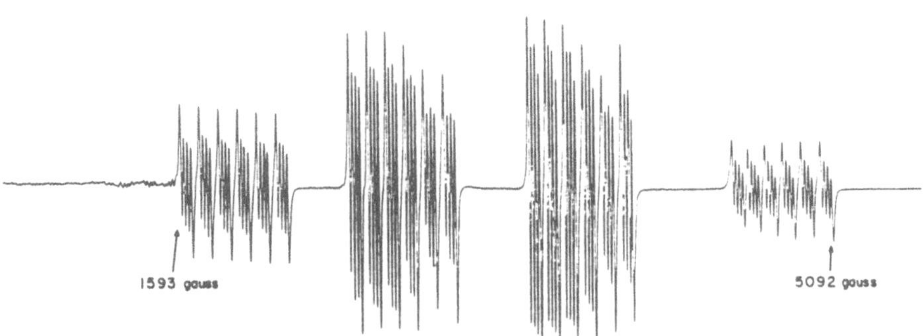

Fig. 27. EPR spectrum of Mn(II)-doped bis[μ-pyridine N-oxide dichloroaquocopper(II)] at 77 K (from Ref. 100)

To finish with this example, it is worthwhile noticing that the Cu(II)Mn(II) pair is isoelectronic with the Cu(II)Fe(III) pairs discussed in Sects. 4.4 and 5.1. These latter species, however, were EPR silent owing likely to a large local anisotropy $\mathbf{D}_{Fe}$, which produced a zero field splitting within the S = 2 state much stronger than in Cu(II)Mn(II).

7.2 *Role of the Number of Unpaired Electrons*

Our second example concerns a series of bimetallic complexes studied by Gagné, Hendrickson et al.[86,103], in which one of the interacting ions remains constant, namely copper(II), and the other one varies from copper(II) to manganese(II). The general formula is $CuBLCl_2$, xH_2O where LH_2 is the binucleating Robson ligand[87] and B = Cu, Ni, Co, Fe and Mn. This series is particularly well adapted for studying the role of the number of unpaired electrons on a center since, along the whole series, the geometry is supposed to be constant and moreover, in each complex, the environments of the two metal ions are identical. The likely structure is as shown in Fig. 28 with the metal ions in square pyramidal surroundings. For all the complexes of the series, the J values characterising the magnitude of the isotropic interaction were deduced from the magnetic data.

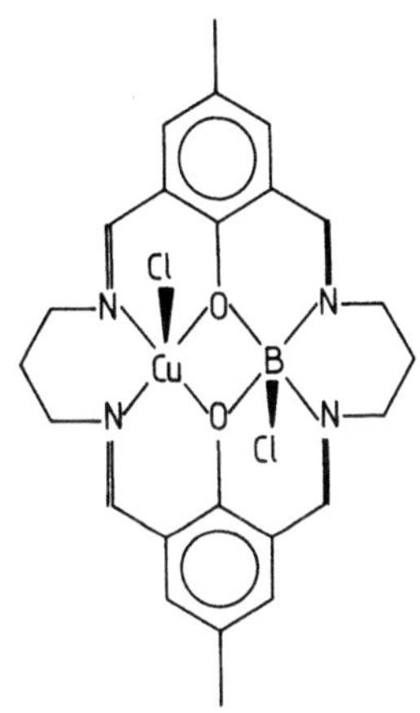

Fig. 28. Likely structure of $CuBLCl_2 \cdot xH_2O$. LH_2 is the binucleating Robson ligand and B = Cu, Ni, Co, Fe and Mn

We do not detail here the fitting procedures and we restrict ourselves to examining the values that were obtained:

	$-J/cm^{-1}$	$-n_An_B J/cm^{-1}$
CuCu	558	558
CuNi	206	412
CuCo	162	486
CuFe	142	568
CuMn	60	300

We must notice, however, that in the case of CuCo and CuFe, the B ion is orbitally degenerate, so that the J value may be less reliable than for the other three compounds.

From these values, there appears a clear trend of a decrease in $|J|$ as the number of unpaired electrons on a center increases. Let us try to rationalize this trend. We have seen in Sect. 3.2 that J can be expressed (relation (6)) as a sum of $J_{\mu\nu}$ contributions involving pairs of magnetic orbitals. This relation (6) shows that the magnitude of the isotropic interaction phenomenon is not expressed by J but by $n_A n_B$ J, which has also been tabulated above. The variation of this quantity along the series apparently does not reveal any clear trend; however our feeling is that the attention should be focused on the three compounds for which there is no problem of orbital degeneracy. When doing that, one sees that not only $|J|$ but also $n_A n_B |J|$ decreases when the number of unpaired electrons increases. This correlation is likely due to the fact that when n_B increases, n_A being equal to 1, the number of ferromagnetic terms $J_{\mu\nu}$ in (6) involving pairs of orthogonal magnetic orbitals increases and the magnitude of the antiferromagnetic interaction is reduced. We think that this correlation is quite general for binuclear complexes of similar symmetry. This does not mean that otherwise a $J_{\mu\nu}$ component for a given pair of magnetic orbitals is constant whatever the nature of the interacting ions may be, provided that the geometry is retained. For instance the $J_{b_1b_1}$ component involving in-plane magnetic orbitals (see Figs. 7 and 18) is likely more antiferromagnetic in the CuCu complex than in the CuNi or CuMn ones. The Cu–O bond lengths in the basal plane are systematically shorter than the Ni–O and Mn–O bond lengths (roughly 1.90 and 2.02 Å respectively), so that the delocalization of the b_1 magnetic orbital toward the bridging ligands could be more prononced with copper than with nickel or manganese. Moreover, we have already pointed out that $J_{b_1b_1}$ is very sensitive to small structural changes such as a modification of the bridging angle (see Sect. 4.3). In other words, a $J_{\mu\nu}$ component is certainly not transferable from one compound to another one. The channel model proposed by Eremin and Rakitin[104, 105], that implicates such a transferability, may have some validity for very rigid structures like those encountered in classical extended lattices (oxides or fluorides) but is irrelevant for the flexible structures of the inorganic chemistry.

7.3 Cobalt-Substituted Superoxide Dismutase

Our last example concerns a nice study of the magnetic susceptibility of the cobalt(II) substituted superoxide dismutase carried out by Morgenstern-Badarau et al.[106]. In a certain sense, this study may be considered as exemplary of what has to be done in magneto-biochemistry. The problem is then to extract a small paramagnetic contribution due to the metal sites from an overall high diamagnetism due to the organic core of the compound. This requires to work with biological materials of high degree of purity and homogeneity and to use a magnetometer of high sensitivity like the Superconducting Quantum Interference Device (SQUID) equipment. But even when these conditions are met, the determination of the large diamagnetic correction remains a crucial problem. The only fully satisfying way to solve it is to study independently the diamagnetic susceptibility of the appropriate apoprotein. In the present case, this has been possible.

The crystal structure of the superoxide dismutase, noted Cu_2Zn_2SOD, has been solved[107–108]. The protein contains two identical Cu(II)Zn(II) pairs per functioning unit.

In each of them, Cu(II) is in square pyramidal surroundings and Zn(II) in tetrahedral surroundings. The metal centers are bridged by the imidazolato group of a histidine. Zn(II) can be substituted by Co(II) to give Cu_2Co_2SOD. The derivatives E_2Co_2SOD and E_2E_2SOD where E symbolizes an empty site have also been prepared. E_2E_2SOD provides the diamagnetic correction of $-1.43 \pm 0.03 \times 10^{-6}$ cm^3g^{-1}. Cu_2Zn_2SOD gives the magnetic behavior of the copper(II) monomeric fragment, E_2Co_2SOD that of the cobalt(II) monomeric fragment and Cu_2Co_2SOD finally that of the Cu(II)Co(II) pair. Cobalt(II) in tetrahedral surroundings has a $^4A_2$ non orbitally degenerate local ground state. Its interaction with copper(II) gives rise to spin triplet and spin quartet low lying states, separated by 2 J. The decrease of $\chi_M T$ upon cooling indicates that the triplet is the ground state, the interaction being antiferromagnetic. This triplet may be split in zero field. If this splitting is assumed to be axial, the magnetic susceptibility is expressed as:

$$\chi_{/\!/} = \frac{2\,N\beta^2}{kT}\,[g_1^2 \exp(-D/kT) + 5\,g_2^2 \exp(2\,J/kT)]/[1 + 2\exp(-D/kT) + 5\exp(2\,J/kT)]$$

$$\chi_{\perp} = 2\,N\beta^2 \left\{\frac{g_1^2}{D}\,[1 - \exp(-D/kT)] + \frac{5\,g_2^2}{kT}\exp(2\,J/kT)\right\} / [1 + 2\exp(-D/kT) + 5\exp(2\,J/kT)] \qquad (52)$$

To derive Eq. (52), it was assumed, as usual, that J was the leading term in the Hamiltonian (2). The $|0\rangle$ level arising from the triplet state has been taken as the origin of the energies. $D = 3\,D_{1zz}/2$ is the axial zero field splitting parameter for the triplet state. As for $\mathbf{g}_1$, $\mathbf{g}_2$ and $\mathbf{D}_1$, they are expressed as:

$$\mathbf{g}_1 = (5\,\mathbf{g}_{Co} - \mathbf{g}_{Cu})/4$$
$$\mathbf{g}_2 = (3\,\mathbf{g}_{Co} + \mathbf{g}_{Cu})/4 \qquad (53)$$

$$\mathbf{D}_1 = 3\,\mathbf{D}_{Co}/2 - \mathbf{D}_{CuCo}/4 \qquad (54)$$

The fitting of the $\chi_M T$ versus T plot (χ_M being the average magnetic susceptibility) leads to the spectrum of the low lying states of Fig. 29. The triplet-quintet energy gap is $2\,J = -33$ cm^{-1}. As expected Cu_2Co_2SOD is EPR silent. To our knowledge, the cobalt-substituted superoxide dismutase provides the only reported example of interaction between copper(II) and non orbitally degenerate cobalt(II).

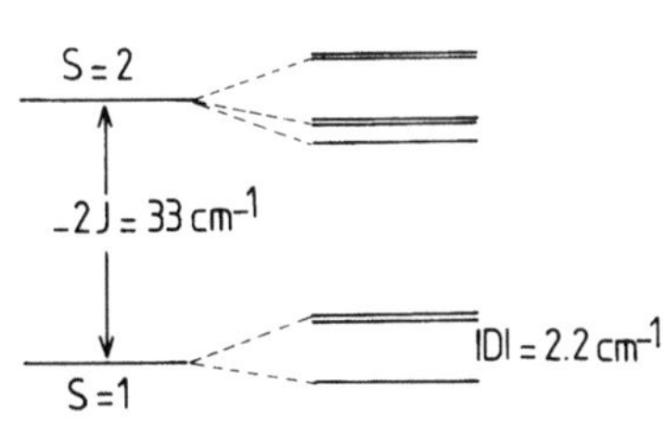

Fig. 29. Spectrum of the low lying states in cobalt(II)-substituted superoxide dismutase (from Ref. 106)

8 Heterobimetallic Complexes with an Orbitally Degenerate Ion

In spite of a fairly large number of works devoted to this subject, the problem of the interaction between orbitally degenerate ions remains an essentially unsolved problem, in the sense where there is no widely accepted model[109–119]. To a large extent each interaction problem involving an orbital degeneracy has been treated as a specific problem with a specific model. In this section, we restrict ourselves to point out the nature of the difficulties encountered with orbital degeneracy and to present some simple ideas allowing to solve them, at least in part.

8.1 *Cu(II)Co(II) in $CuCo(fsa)_2en(H_2O)_2 \cdot H_2O$*

To start with, we consider the complex $CuCo(fsa)_2en(H_2O)_2 \cdot H_2O$, noted [CuCo], of which the structure is very similar to that shown in Fig. 15, Co(II) replacing Ni(II) in the octahedral site[85]. This system may be viewed as the simplest one involving an orbital degeneracy since only one ion is orbitally degenerate, the other ion being a spin doublet[120].

In [CuCo], the local ground states are $^2B_1$ for the copper(II) fragment and $^4T_1$ for the cobalt(II) fragment. Refering to the C_{2v} site symmetry, $^4T_1$ is split into $^4A_2 + {}^4B_1 + {}^4B_2$, each of them being then split into two Kramers doublets by spin orbit coupling. If the distortion with respect to the pure octahedron is not too pronounced, this latter splitting within the quartet local states may be much larger than the local anisotropy of the non orbitally degenerate ions. It follows that the approach utilized in the preceding sections is not valid anymore. In particular the isotropic interaction cannot be properly described by a unique parameter J. Of course, in the opposite case of an extremely strong distortion around Co(II), one of the quartet local states would be highly stabilized with respect to the other two and the orbital degeneracy would vanish. In such a border line case, an expression like (52) would be a valid description of the magnetic susceptibility of the Cu(II)Co(II) pair, provided that the interaction is antiferromagnetic and sufficiently large. But, let us turn back to the situation of orbital quasi degeneracy and let us present the broad outlines of the simplest approach, in principle appropriate for any A − B case. The Hamiltonian in zero field is written as

$$\mathcal{H} = \mathcal{H}_A + \mathcal{H}_B + \mathcal{H}_{AB} \tag{55}$$

$\mathcal{H}_A$ (or $\mathcal{H}_B$) explicitely takes into account the distortion in the monomeric fragment assumed to be orbitally degenerate as well as the local spin orbit coupling. In other words, the eigenstates of $\mathcal{H}_A$ are the local states arising from the orbitally degenerate ground state, for instance six Kramers doublets for Co(II) in distorted octahedral surroundings. $\mathcal{H}_{AB}$ is defined as

$$\mathcal{H}_{AB} = -\hat{J}\hat{S}_A \cdot \hat{S}_B \tag{56}$$

where $\hat{S}_A$ and $\hat{S}_B$ are the local spin operators and $\hat{J}$ an orbital operator acting on the pair functions obtained as products of the local functions and transforming as the irreducible representations Γ of the appropriate symmetry group, according to:

$$\langle \Gamma | \hat{J} | \Gamma \rangle = J_\Gamma \delta_{\Gamma\Gamma'} \tag{57}$$

The Zeeman perturbation is finally expressed as the sum of the local contributions:

$$\mathcal{H}_Z = \mathcal{H}_{ZA} + \mathcal{H}_{ZB} \tag{58}$$

with:

$$\mathcal{H}_{ZA} = \beta \vec{H} \cdot (k_A \hat{L}_A + g_e \hat{S}_A) \tag{59}$$

and an identical expression for $\mathcal{H}_{ZB} \cdot \hat{L}_A$ and $\hat{L}_B$ are angular momentum operators, k_A and k_B covalence reduction factors assumed here to be isotropic and g_e is the gyromagnetic ratio for the free electron. In the case where one of the interacting ions, say B, is an orbital singlet, $\mathcal{H}_{ZB}$ becomes:

$$\mathcal{H}_{ZB} = \beta \vec{H} \cdot \mathbf{g}_B \cdot \hat{S}_B \tag{60}$$

The three principal magnetic susceptibilities χ_x, χ_y and χ_z can then be calculated through the Van Vleck equation, which requires the eigenvalues and eigenfunctions of $\mathcal{H}$ defined in Eq. (55), and the first and second order Zeeman coefficients.

Let us apply this model to the [CuCo] case. The interaction $\mathcal{H}_{CuCo}$ between the ${}^2B_1$ and ${}^4T_1$ local states may be described, using a symbolic notation, by:

$$
{}^2B_1 \leftrightarrow {}^4T_1 \quad
\begin{array}{l}
\nearrow\ {}^2B_1 \leftrightarrow {}^4A_2 \rightarrow {}^3B_2 + {}^5B_2 \\
\longrightarrow\ {}^2B_1 \leftrightarrow {}^4B_1 \rightarrow {}^3A_1 + {}^5A_1 \\
\searrow\ {}^2B_1 \leftrightarrow {}^4B_2 \rightarrow {}^3A_2 + {}^5A_2
\end{array}
\tag{61}
$$

Therefore, Γ in (57) can be A_1, A_2 and B_2, and $2\,J_{A_1}$, $2\,J_{A_2}$ and $2\,J_{B_2}$ are the energy gaps between the ${}^3A_1 - {}^5A_1$, ${}^3A_2 - {}^5A_2$ and ${}^3B_2 - {}^5B_2$ pair states, respectively. These J_Γ parameters can be easily decomposed in sums of $J_{\mu\nu}$ components involving pairs of interacting orbitals. This is perhaps the most heuristic aspect of this approach; *it leads to the same orbital interpretation of the interaction parameters as in the case where there is no orbital degeneracy*. For [CuCo], there is one unpaired electron around copper in a b_1 magnetic orbital and three unpaired electrons around cobalt that can occupy the five magnetic orbitals $a_1(z^2)$, $a_1'(x^2 - y^2)$, $a_2(yz)$, $b_1(xy)$ and $b_2(zx)$. The orbital parts of the wavefunctions associated with ${}^4A_2$, ${}^4B_1$ and ${}^4B_2$ local states are:

$$
\begin{aligned}
{}^4A_2 &= \alpha(a_1b_1b_2) - (1 - \alpha^2)^{1/2}(a_1a_1'a_2) \\
{}^4B_1 &= \alpha(a_1b_1a_1') - (1 - \alpha^2)^{1/2}(a_1a_2b_2) \\
{}^4B_2 &= \alpha(a_1b_1a_2) - (1 - \alpha^2)^{1/2}(a_1a_1'b_2)
\end{aligned}
\tag{62}
$$

α varies from $2\sqrt{5}/5$ in the weak field limit to 1 in the high field limit[121]. From Eq. (62), it is straightforward to obtain:

$$J_{A_1} = 2/3[\alpha^2(J_{b_1a_1} + J_{b_1b_1} + J_{b_1a_1'}) + (1-\alpha^2)(J_{b_1a_1} + J_{b_1a_2} + J_{b_1b_2})]$$
$$J_{A_2} = 2/3[\alpha^2(J_{b_1a_1} + J_{b_1b_1} + J_{b_1a_2}) + (1-\alpha^2)(J_{b_1a_1} + J_{b_1a_1'} + J_{b_1b_2})] \quad (63)$$
$$J_{B_2} = 2/3[\alpha^2(J_{b_1a_1} + J_{b_1b_1} + J_{b_1b_2}) + (1-\alpha^2)(J_{b_1a_1} + J_{b_1a_1'} + J_{b_1a_2})]$$

It should be noticed that all the $J_{\mu\nu}$ components occuring in Eq. (63) are positive (ferromagnetic) but $J_{b_1b_1}$. It can be also noticed that the coefficient of $J_{b_1b_1}$ is equal to 2 $\alpha^2/3$ for the three J_Γ parameters. It follows that, if we assume that this antiferromagnetic contribution $J_{b_1b_1}$ is dominant, the J_{A_1}, J_{A_2} and J_{B_2} parameters should have values close to each other.

For the calculation, the basis set is made out of the 24 functions $|M_L, S, M_S\rangle$. (M_L (= 0, ± 1) is the quantum number associated with the projection of $\hat{L}_{Co}$ along the z axis. S (= 1 and 2) and M_S are the spin quantum numbers for [CuCo].) $\mathcal{H}_{Co}$ is written as:

$$\mathcal{H}_{Co} = \hat{L}_{Co} \cdot \mathbf{D} \cdot \hat{L}_{Co} + k_{Co}\lambda\hat{L}_{Co} \cdot \hat{S}_{Co} \quad (64)$$

D is a tensor describing the local distortion around Co(II) and λ the spin orbit coupling coefficient for the ^4F free ion ground state. In fact, it is convenient to use the T ≡ P analogy. This requires replacing $\hat{L}_{Co}$ both in $\mathcal{H}_{Z_{Co}}$ and the spin orbit coupling term of Eq.(64) by $-3\,\hat{L}_{Co}/2$ to account for the fact that the real angular momentum of the $^4T_1$ state is equal to that of the ^4P state multiplied by $-3/2$[121].

This model permits a satisfactory interpretation of the average magnetic susceptibility of [CuCo]. For the sake of simplicity, it was assumed that **D** was axial. The three J parameters are then found close to each other with $J_{A_1} \sim J_{A_2} \sim J_{B_2} = -62 \pm 10\ cm^{-1}$. This means that for any Γ, the triplet pair state is lower in energy than the quintet, which is not surprising since the strongly antiferromagnetic component $J_{b_1b_1}$ is present in the three J_Γ parameters. The fitting of the magnetic data leads to the spectrum of the eight thermally populated pair levels shown in Fig. 30. This spectrum must be considered as indicative and the signification of the agreement between experimental and calculated magnetic data must not be surestimated. Before reaching a definitive conclusion concerning the actual validity of the model, it is absolutely necessary to apply it to the principal susceptibilities. Unfortunatly, it has not been possible so far to grow single crystals of [CuCo] large enough for magnetic anisotropy measurements.

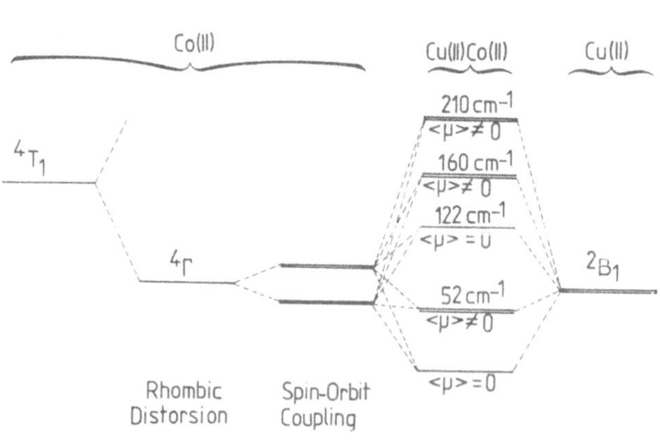

Fig. 30. Spectrum of the thermally populated low lying states in CuCo(fsa)$_2$en(H_2O)$_2$ · H_2O

8.2 Dilute Ni(II)Co(II) Pairs

The model described in the preceding section has also been utilized for interpreting the EPR spectra associated with the ground Kramers doublet of two magnetically dilute Ni(II)Co(II) pairs. In one case[96], the host lattice is $M_2(trik)(py)_4$, M = Ni or Co, of which the structure has been schematized in Sect. 6.5; in the other[97], the host lattice is $[M_2(phph)_2(H_2O)_4]Cl_4 \cdot H_2O$, M = Ni or Co and phph = 1,4-dihydrazinophtalazine, with the structure schematized as:

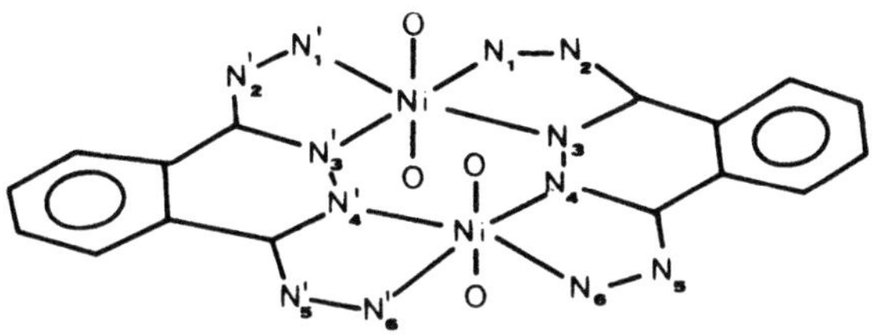

For both systems, site and molecular symmetry is C_{2v}. $\mathcal{H}_{Co}$ is expressed as in Eq. (64) and $\mathcal{H}_{Ni}$ that splits in zero field the non orbitally degenerate $^3B_1$ local state as:

$$\mathcal{H}_{Ni} = \hat{S}_{Ni} \cdot \mathbf{D}_{Ni} \cdot \hat{S}_{Ni} \tag{65}$$

$\mathcal{H}_{NiCo}$ acts according to the scheme:

$$^3B_1 \leftrightarrow {}^4T_1 \begin{cases} \nearrow {}^3B_1 \leftrightarrow {}^4A_2 \rightarrow {}^2B_2 + {}^4B_2 + {}^6B_2;\ J_{B_2} \\ \rightarrow {}^3B_1 \leftrightarrow {}^4B_1 \rightarrow {}^2A_1 + {}^4A_1 + {}^6A_1;\ J_{A_1} \\ \searrow {}^3B_1 \leftrightarrow {}^4B_2 \rightarrow {}^2A_2 + {}^4A_2 + {}^6A_2;\ J_{A_2} \end{cases} \tag{66}$$

where J_Γ, $\Gamma = A_1$, A_2 and B_2, measures the energy gaps between doublet and quartet pair states ($3\ J_\Gamma/2$), and doublet and sextet pair states ($4\ J_\Gamma$) of orbital symmetry Γ. The method used by Gatteschi et al. to fit the EPR spectra was the following[96, 97]: the parameters of $\mathcal{H}_{Ni}$ and $\mathcal{H}_{Co}$ were first determined from experimental data like EPR spectra of the Zn(II)Co(II) and Ni(II)Cu(II) pairs and single crystal polarized electronic spectra, then the J_Γ interaction parameters were calculated by least square fitting. For both systems, a satisfying agreement was reached with reasonable values of J_{A_1}, J_{A_2} and J_{B_2}. A typical result, that of Ni(II)Co(II) in $M_2(trik)(pyr)_4$ is $J_{A_1} = -4 \pm 4\ cm^{-1}$; $J_{A_2} \simeq J_{B_2} = -45 \pm 5\ cm^{-1}$, leading to $g_x = 0.46$, $g_y = 1.20$ and $g_z = 2.10$ while the experimental values are $g_x = 0.3$, $g_y = 1.2$ and $g_z = 2.1$. Once again, the satisfying experiment-theory agreement should be considered as an indication rather than a definitive proof of the validity of the model.

8.3 Cu(II)-Low Spin Fe(III) Pair

Gatteschi et al. have proposed another approach, somewhat different conceptually from that presented just above, in order to describe the interaction between copper(II) and low spin iron(III) in the cyanide form of cytochrome oxidase[122] (see Sect. 5.2). This model does not take into account the full orbital degeneracy of the $^2T_2$ state for Fe(III) but only the ground Kramers doublet of the heme fragment. This is possible because the local functions φ and φ^- associated with this Kramers doublet, i.e. the degree of mixing of the xy-, yz- and zx-type magnetic orbitals, can be determined from the EPR spectrum of the heme fragment according to[123, 124]:

$$\varphi = a\phi_{yz} - ib\phi_{zx} - c\bar{\phi}_{xy}$$
$$\bar{\varphi} = -a\bar{\phi}_{yz} - ib\bar{\phi}_{zx} - c\phi_{xy} \quad (67)$$

with:

$$a = (g_z + g_y)/h$$
$$b = (g_z - g_x)/h \quad (68)$$
$$c = (g_y - g_x)/h$$

and:

$$h = [8(g_z + g_y - g_x)]^{1/2} \quad (69)$$

The basis set is then made out of the four functions constructed by the product of the local ground state functions and the energies of the pair levels are calculated applying an effective interaction Hamiltonian $\mathcal{H}_{CuFe}$, of which the matrix elements are expressed in function of two-electron exchange integrals J_{ij} involving the magnetic orbital around copper(II) and two magnetic orbitals ϕ_i and ϕ_j around iron(III). If the symmetry is that of Fig. 13, with CN^- instead of Cl^- (i.e. C_{4v}), all the J_{ij} integrals with $i \neq j$ vanish, but if the Cu(II) Fe(III) pair has no overall symmetry condition, all the J_{ij} integrals may be different from zero. The interesting point of this model is that the matrix elements of $\mathcal{H}_{CuFe}$ using the four spin orbital pair functions as basis set can be equated to the matrix elements of the phenomenological spin Hamiltonian:

$$\mathcal{H} = -J\hat{S}_{Cu} \cdot \hat{S}_{Fe} + \hat{S}_{Cu} \cdot \mathbf{D} \cdot \hat{S}_{Fe} + \vec{d} \cdot \hat{S}_{Cu} \wedge \hat{S}_{Fe} \quad (70)$$

using the four $|M_{S_{Cu}}, M_{S_{Fe}}\rangle$ spin functions as basis set. This provides relations between the parameters of Eq. (70) on the one hand, the mixing coefficients in Eq. (67) and the J_{ij} integrals on the other hand. These relations as tabulated by Gatteschi et al. are:

$$J = 2a^2J_{11}/3 - 2b^2J_{22} + 2c^2J_{33}/3$$
$$D_{xx} = 8a^2J_{11}/3 + 4c^2J_{33}/3$$
$$D_{yy} = 4a^2J_{11}/3$$
$$D_{zz} = 4a^2J_{11}/3 - 8c^2J_{33}/3$$
$$D_{xy} = D_{yz} = 0 \quad (71)$$
$$D_{zx} = 4acJ_{31}$$
$$d_x = 4bcJ_{23}$$
$$d_y = 0$$
$$d_z = -4abJ_{12}$$

The indices 1, 2 and 3 stand for yz, zx and xy respectively. This model may be considered as intermediate in sophistication between a model using the 12 functions associated with the $^2B_1 \leftrightarrow {}^2T_2$ interaction as basis set and the model defined by the spin Hamiltonian (70) where the physical meaning of the parameters does not appear. The information lost by working with a basis set limited to four pair functions is recovered by injecting the EPR data concerning the heme fragment, which allows the calculation of the orbital structure of the local ground Kramers doublet.

Another formulation of this model using the second quantization technique has also been proposed[122]. The basic ideas as well as the main conclusions are unchanged.

9 Heterotrinuclear Complexes

Relatively few heterotrinuclear complexes with magnetic centers have been characterized and investigated in a reliable fashion. In this section, we again selected some examples which appeared to us particularly significant. Other examples will be presented in the next section.

9.1 $\{[Cu(Mesalen)]_2VO\}(ClO_4)_2$

Many trinuclear species resulting from the condensation of tetradentate Schiff base copper(II) complexes with various divalent ions were synthesized by Sinn et al. almost two decades ago[125–127]. Preliminary magnetic studies were also reported. The same kind of synthesis has been used by Gatteschi et al. to obtain $\{[Cu(Mesalen)]_2VO\}(ClO_4)_2$, with Mesalen = N,N′-bis(methylsalicylaldiminato)ethylenediamine[128]. The structure of the compound is shown in Fig. 31. VO(II) is bridged to two Cu(II) ions through the phenolic

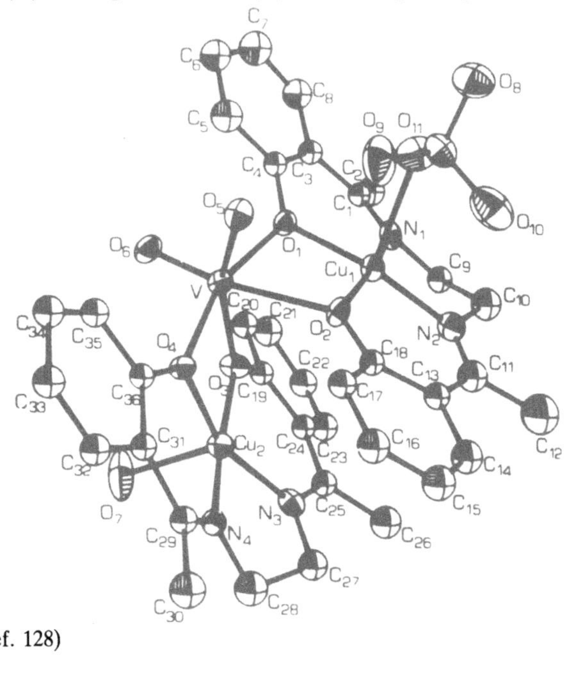

Fig. 31. Crystal structure of $\{[Cu(Mesalen)]_2VO\}(ClO_4)_2$ (from Ref. 128)

oxygen atoms of the tetradentate Schiff base, but the two CuO_2V bridging networks are quite different. In Cu1–O1–O2–V, O1 and O2 are in the basal plane of copper in square pyramidal surroundings; if O1 is also in the basal plane of the vanadyl ion with $V - O1 = 1.99$ Å, O2 occupies the apical position, opposite to that of the vanadyl oxygen atom with $V - O2 = 2.35$ Å. The other bridging network, on the other hand, is very similar to that encountered in $CuVO(fsa)_2en \cdot CH_3OH$ (see Sect. 4.1), the two basal planes around copper and vanadium respectively sharing the O3–O4 edge. Each magnetic center has a unique unpaired electron and the magnetic orbitals may be schematized as shown in Fig. 32. In the former bridging network, the magnetic orbitals centered on Cu1 and V may give a non zero overlap density around O1 but no overlap density around O2. The orientations of the 2p orbitals on O1 involved in ϕ_{Cu1} and ϕ_{VO} do not allow an accidental orthogonality so that a weak antiferromagnetic interaction between Cu1 and VO may be expected ($J_1 < 0$). As for the latter bridging network, it has all the symmetry requirements leading to a strict orthogonality of ϕ_{Cu2} and ϕ_{VO} and the interaction between the metal centers is certainly ferromagnetic with J_2 of the order of 100 cm^{-1} (see Sect. 4.1). Finally the interaction between the terminal copper atoms through the VO_6 chromophore is likely very weak ($j \sim 0$).

The spin Hamiltonian appropriate to this problem is:

$$\mathcal{H} = - J_1\hat{S}_{Cu1} \cdot \hat{S}_{VO} - J_2\hat{S}_{Cu2} \cdot \hat{S}_{VO} - j\hat{S}_{Cu1} \cdot \hat{S}_{Cu2} \qquad (72)$$

where we neglected the anisotropic and antisymmetric interaction terms. The low lying states are a quartet and two doublets. Since the relative energies of those states depend on only two energy gaps, the three parameters of Eq. (72) cannot be determined univocally from the magnetic susceptibility data and a simplifying hypothesis must be made; the

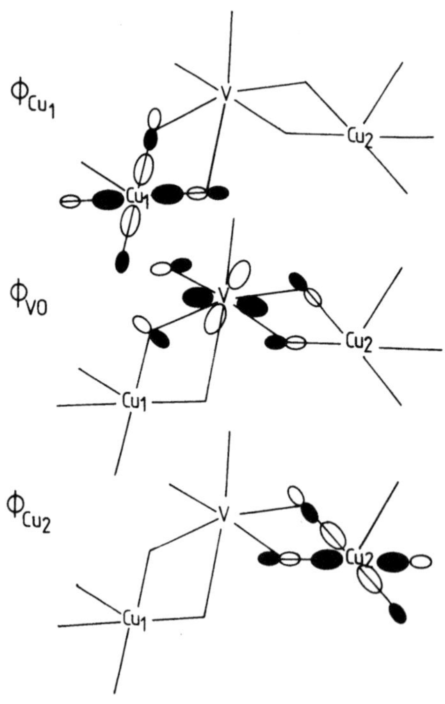

Fig. 32. Schematic representation of the magnetic orbitals in $\{[Cu(Mesalen)]_2VO\}(ClO_4)_2$

most obvious one consists of neglecting j. Whereas the vector model can be utilized to calculate the relative energies of the states when J_1 and J_2 are equal, even with $j \neq 0$, in the present case it is necessary to diagonalize Eq. (72) using the spin functions as basis set.

($\alpha\alpha\alpha$) is the component $M_S = 3/2$ of the quartet and its energy is easily found as $-(J_1 + J_2)/4$. The matrix associated with Eq. (72) in the subspace $M_S = 1/2$ is:

($\alpha\alpha\beta$)	($\alpha\beta\alpha$)	($\beta\alpha\alpha$)	
$(-J_1 + J_2)/4$	$-J_2/2$	0	
	$(J_1 + J_2)/4$	$-J_1/2$	(73)
		$(J_1 - J_2)/4$	

of which the eigenvalues are $-(J_1 + J_2)/4$ and $(J_1 + J_2)/4 \pm (\sqrt{J_1^2 + J_2^2 - J_1J_2})/2$. By changing the energy origin, the relative energies E(S) may be written as:

$$\begin{aligned} E_1(1/2) &= 0 \\ E_2(1/2) &= -\sqrt{J_1^2 + J_2^2 - J_1J_2} = \Delta_1 \\ E(3/2) &= -(J_1 + J_2 + \sqrt{J_1^2 + J_2^2 - J_1J_2})/2 = \Delta_2 \end{aligned} \tag{74}$$

and the magnetic susceptibility as:

$$\chi_M T = \frac{N\beta^2}{4\,k} \; \frac{g_{1/2(1)}^2 + g_{1/2(2)}^2 \exp(-\Delta_1/kT) + 10\, g_{3/2}^2 \exp(-\Delta_2/kT)}{1 + \exp(-\Delta_1/kT) + 2\exp(-\Delta_2/kT)} \tag{75}$$

$g_{1/2(1)}$ and $g_{1/2(2)}$ refer here to the doublet states of energies $E_1(1/2)$ and $E_2(1/2)$ respectively. In absence of symmetry, the g-factors associated to the triad states cannot be easily related to the local g-factors. The magnetic susceptibility data can be fitted with values of Δ_1 and Δ_2 close to each other and negative ($\Delta_1 \sim \Delta_2 \sim -90\ \mathrm{cm}^{-1}$). This means that one of the doublet states and the quartet state are almost accidentally degenerate ground states, with the second doublet state being located at roughly 90 cm^{-1} above. This spectrum is compatible with J_1 close to zero and J_2 close to 90 cm^{-1}. The powder EPR spectra also suggest that both a doublet and a quartet states are populated at 4.2 K.

9.2 {[Cu(Mesalen)]$_2$Fe(acac)}(NO$_3$)$_2$

Another interesting heterotrinuclear complex synthesized from Cu(Mesalen) is {[Cu(Mesalen)]$_2$Fe(acac)}(NO_3)$_2$ with acac = acetylacetonato, containing the Cu(II)Fe(III)Cu(II) triad. It has been studied by Morgenstern-Badarau and Wickman[129]. Its crystal structure has not yet been solved but it is most likely as schematized in Fig. 33 with two equivalent CuO_2Fe bridging networks. Fe(III) is in a local high-spin $S_{Fe} = 5/2$ state and the spin Hamiltonian used to interpret the magnetic data is:

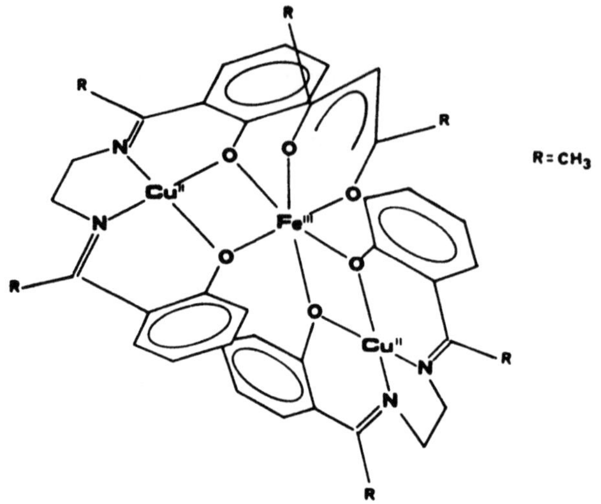

Fig. 33. Structure of $\{[Cu(Mesalen)]_2Fe(acac)\}^{2+}$ (from Ref. 129)

$$\mathscr{H} = -J(\hat{S}_{Cu1} \cdot \hat{S}_{Fe} + \hat{S}_{Cu2} \cdot \hat{S}_{Fe}) + \hat{S}_{Fe} \cdot \mathbf{D}_{Fe} \cdot \hat{S}_{Fe} + \hat{S}_{Cu1} \cdot \mathbf{D}_{Cu1Fe} \cdot \hat{S}_{Fe} + \hat{S}_{Cu2} \cdot \mathbf{D}_{Cu2Fe} \cdot \hat{S}_{Fe} \tag{76}$$

where the interaction between the terminal Cu(II) ions has been neglected. The relative energies E(S, S′) of the triad states arising from the isotropic interaction can be easily calculated using the vector model as:

$$\begin{aligned} E(3/2,\ 1) &= 0 \\ E(5/2,\ 1) &= -5\,J/2 \\ E(5/2,\ 0) &= -7\,J/2 \\ E(7/2,\ 1) &= -6\,J \end{aligned} \tag{77}$$

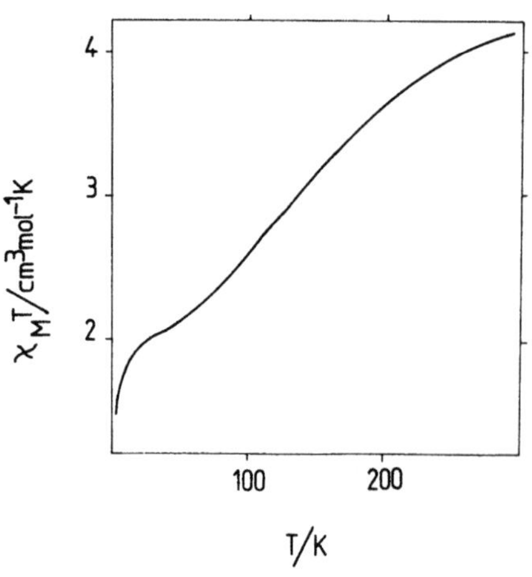

Fig. 34. χ_MT versus T plot for $\{[Cu(Mesalen)]_2Fe(acac)\}(NO_3)_2$ (from Ref. 129)

with:

$$\hat{S}' = \hat{S}_{Cu1} + \hat{S}_{Cu2}$$
$$\hat{S} = \hat{S}' + \hat{S}_{Fe} \qquad (78)$$

Upon cooling, $\chi_M T$ decreases to a plateau characterized by $\chi_M T = 2.03\ cm^3\ mol^{-1}\ K$, then decreases again below 25 K. This behavior (Fig. 34) indicates that the Cu(II)Fe(III) interaction is antiferromagnetic with a quartet ground state split in zero field into two Kramers doublets. Assuming that the zero field splitting within the ground state is axial, the magnetic susceptibilities are:

$$\chi_{/\!/} = \frac{N\beta^2}{4\ kT}\{g_{3/2}^2[\exp(D/kT) + 9\exp(-D/kT)] + 35\ g_{5/2(1)}^2 \exp(5\ J/2\ kT)$$
$$+ 35\ g_{5/2(0)}^2 \exp(7\ J/2\ kT) + 84\ g_{7/2}^2 \exp(6\ J/kT)\}/[\exp(D/kT)$$
$$+ \exp(-D/kT) + 3\exp(5\ J/2\ kT) + 3\exp(7\ J/2\ kT) + 4\exp(6\ J/kT)]$$

$$\chi_{\perp} = \frac{N\beta^2}{4}\left\{\frac{4\ g_{3/2}^2}{kT}\exp(D/kT) + \frac{3\ g_{3/2}^2}{D}[\exp(D/kT) - \exp(-D/kT)]\right.$$
$$+ \frac{35\ g_{5/2(1)}^2}{kT}\exp(5\ J/2\ kT) + \frac{35\ g_{5/2(0)}^2}{kT}\exp(7\ J/kT)$$
$$\left. + \frac{84\ g_{7/2}^2}{kT}\exp(6\ J/kT)\right\}/[\exp(D/kT) + \exp(-D/kT)$$
$$+ 3\exp(5\ J/2\ kT) + 3\exp(7\ J/2\ kT) + 4\exp(6\ J/kT)] \qquad (79)$$

with $D = 3\ D_{zz}/2$, **D** being the zero field splitting tensor for the quartet state. **D** is related to the tensors appearing in Eq. (76) by:

$$\mathbf{D} = -7(\mathbf{D}_{Cu1Fe} + \mathbf{D}_{Cu2Fe})/30 + 28\ \mathbf{D}_{Fe}/15 \qquad (80)$$

and the relations between molecular tensors $\mathbf{g}_{S(S')}$ and local tensors of the Zeeman perturbation:

$$\mathcal{H}_Z = \beta\vec{H}\cdot(\mathbf{g}_{Cu1}\cdot\hat{S}_{Cu1} + \mathbf{g}_{Fe}\cdot\hat{S}_{Fe} + \mathbf{g}_{Cu2}\cdot\hat{S}_{Cu2}) \qquad (81)$$

are:

$$\mathbf{g}_{3/2} = (7\ \mathbf{g}_{Fe} - \mathbf{g}_{Cu1} - \mathbf{g}_{Cu2})/5$$
$$\mathbf{g}_{5/2(1)} = (31\ \mathbf{g}_{Fe} + 2\ \mathbf{g}_{Cu1} + 2\ \mathbf{g}_{Cu2})/35 \qquad (82)$$
$$\mathbf{g}_{5/2(0)} = \mathbf{g}_{Fe}$$
$$\mathbf{g}_{7/2} = (5\ \mathbf{g}_{Fe} + \mathbf{g}_{Cu1} + \mathbf{g}_{Cu2})/7$$

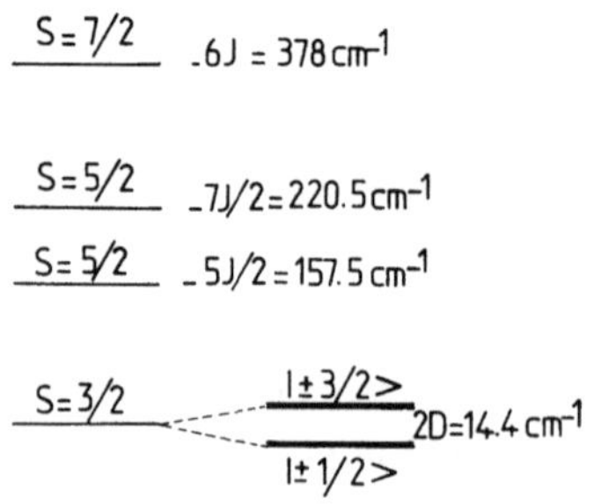

Fig. 35. Spectrum of the low lying states in $\{[Cu(Mesalen)]_2Fe(acac)\}(NO_3)_2$ (from Ref. 129)

The relations (80) and (82) have been derived by Gatteschi and Bencini[27].

The fitting of the magnetic data leads to $J = -63\ cm^{-1}$ and $D = 7.2\ cm^{-1}$ with an average g-value of 2.08, so that the spectrum of the low lying states is as shown in Fig. 35.

The EPR spectrum at 4.2 K agrees with the magnetic data. Its displays the features characteristic of the $|\pm 1/2\rangle$ Kramers doublet arising from the splitting in zero field of the quartet state with $g_1 = 4.68$, $g_2 = 3.26$ and $g_3 = 1.96$. The anisotropy of this spectrum indicates that D is positive, the $|\pm 1/2\rangle$ level being lower than the $|\pm 3/2\rangle$, whereas the sign of D could not be deduced unambiguously from the powder magnetic susceptibility data. On the other hand, the $|\pm 3/2\rangle$ upper Kramers doublet is not detected in EPR, probably because of its anisotropy.

Another very interesting $d^1d^5d^1$ complex has been reported, namely $[(C_5H_5)_2TiCl]_2MnCl_2(thf)_2$ with thf = tetrahydrofuran[130]. Its structure has been solved; two terminal pseudotetrahedral Ti(III) are linked to an octahedral Mn(II) through chloride bridges. Unfortunatly, the magnetic and EPR data are less accurate than for the previous complex. Nevertheless, they show that again the nearest neighbor metal centers are antiferromagnetically coupled with a quartet ground state. The isotropic interaction parameter between Ti(III) and Mn(II) was estimated as $-16.5\ cm^{-1}$.

9.3 $[Cu_2Ni(oxpn)_2(H_2O)_2](ClO_4)_2$

The condensation of two moles of Cu(oxpn) with one mole of Ni(II) perchlorate affords $[Cu_2Ni(oxpn)_2(H_2O)_2](ClO_4)_2$. oxpn is N,N′-bispropylenediamine oxamido. The structure of the trinuclear cation with the Cu(II)Ni(II)Cu(II) triad is schematized in Fig. 36[131].

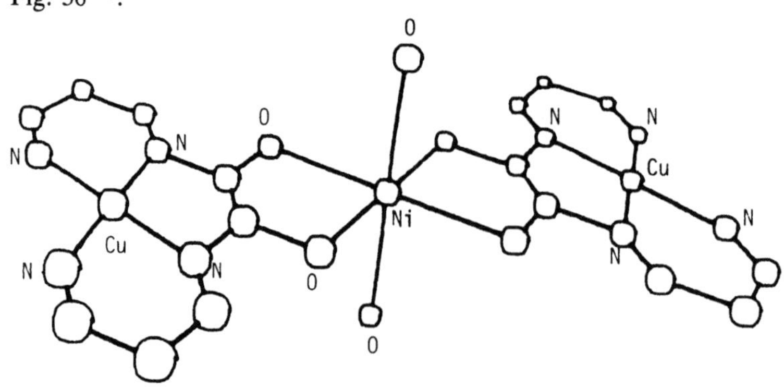

Fig. 36. Structure of $[Cu_2Ni(oxpn)_2(H_2O)_2]^{2+}$ (from Ref. 131)

The spin Hamiltonian in zero field for this compound is written as:

$$\mathcal{H} = -J(\hat{S}_{Cu1} \cdot \hat{S}_{Ni} + \hat{S}_{Cu2} \cdot \hat{S}_{Ni}) - j\hat{S}_{Cu1} \cdot \hat{S}_{Cu2} \tag{83}$$

and the relative energies E(S, S′) of the low lying states are:

$$\begin{aligned} E(0, 1) &= 0 \\ E(1, 1) &= -J \\ E(1, 0) &= -2J + j \\ E(2, 1) &= -3J \end{aligned} \tag{84}$$

with:

$$\begin{aligned} \hat{S}' &= \hat{S}_{Cu1} + \hat{S}_{Cu2} \\ \hat{S} &= \hat{S}' + \hat{S}_{Ni} \end{aligned} \tag{85}$$

Since the ground state of the compound is the singlet (vide infra), it is not necessary to introduce local anisotropy or anisotropic interaction terms in Eq. (83). The Zeeman perturbation is:

$$\mathcal{H}_Z = \beta\vec{H} \cdot (\mathbf{g}_{Cu1} \cdot \hat{S}_{Cu1} + \mathbf{g}_{Ni} \cdot \hat{S}_{Ni} + \mathbf{g}_{Cu2} \cdot \hat{S}_{Cu2}) \tag{86}$$

and the magnetic susceptibility is:

$$\chi_M T = \frac{2N\beta^2}{k} \cdot [g_{1(1)}^2 \exp(J/kT) + g_{1(0)}^2 \exp((2J - j)/kT)$$

$$+ 5 g_2^2 \exp(3J/kT)]/[1 + 3\exp(J/kT) + 3\exp((2J - j)/kT)$$
$$+ 5\exp(3J/kT)] \tag{87}$$

with:

$$\begin{aligned} \mathbf{g}_{1(1)} &= \mathbf{g}_2 = (\mathbf{g}_{Cu1} + \mathbf{g}_{Cu2} + 2\,\mathbf{g}_{Ni})/4 \\ \mathbf{g}_{1(0)} &= \mathbf{g}_{Ni} \end{aligned} \tag{88}$$

The only role of j is to modify slightly E(1, 0), the energies of the other three states remaining unchanged. Its influence on the magnetic properties is almost negligible, so that it cannot be determined by fitting the magnetic data of the Cu(II)Ni(II)Cu(II) species. On the other hand, j can be estimated from the magnetic properties of the Cu(II)Zn(II)Cu(II) species where Zn(II) replaces Ni(II). This compound behaves as a copper(II) binuclear complex with a singlet-triplet energy gap equal to j which lies in the range $0 > j > -1.6\ cm^{-1}$. The magnetic properties of $[Cu_2Ni(oxpn)_2(H_2O)_2](ClO_4)_2$ are shown in Fig. 37, where, to be consistent with the other magnetic curves of this article, we plotted $\chi_M T$ versus T. Upon cooling, $\chi_M T$ decreases continuously and tends to a value

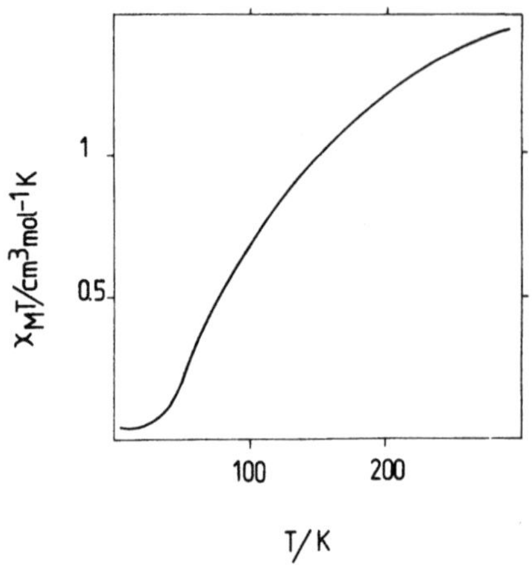

Fig. 37. χ_MT versus T plot for $[Cu_2Ni(oxpn)_2(H_2O)_2](ClO_4)_2$ (from Ref. 131)

very close to zero, confirming that the ground state is the singlet. The χ_M versus T plot exhibits a maximum around 90 K, which is characteristic of a coupled molecular system with a diamagnetic ground state. The fitting of the data leads to $J = -98.7\ cm^{-1}$ with an average g-value of 2.12 and a spectrum of the low lying states is depicted in Fig. 38, where we indicated the orbital symmetry of the states, assuming a D_{2h} molecular symmetry.

The most interesting result for this complex is the large magnitude of the antiferromagnetic interaction between metal ions separated by about 5.1 Å. In other words, the oxamido bis bidentate ligand has quite a remarkable ability to transmit the electronic effects between metal ions far away from each other. This phenomenon has recently been reviewed[132] and we do not intend to discuss it again here. We simply recall that the efficiency of the bis bidentate ligands like those shown below, with X, Y = O, S or NR,

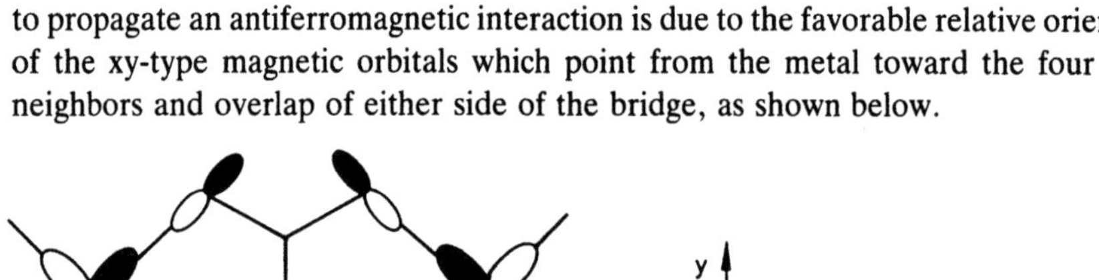

to propagate an antiferromagnetic interaction is due to the favorable relative orientations of the xy-type magnetic orbitals which point from the metal toward the four nearest neighbors and overlap of either side of the bridge, as shown below.

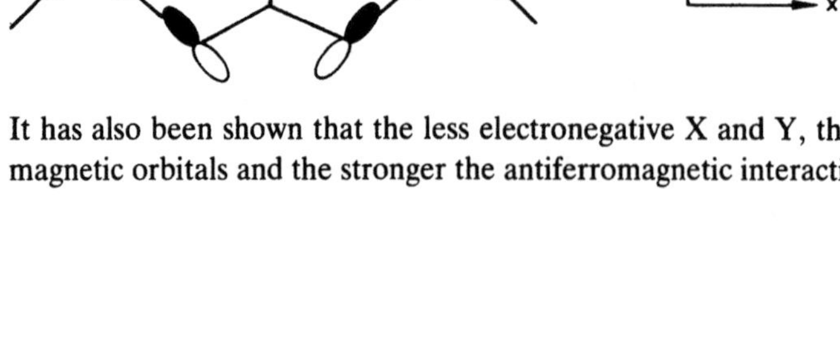

It has also been shown that the less electronegative X and Y, the more delocalized the magnetic orbitals and the stronger the antiferromagnetic interaction[132, 133]. So, the mag-

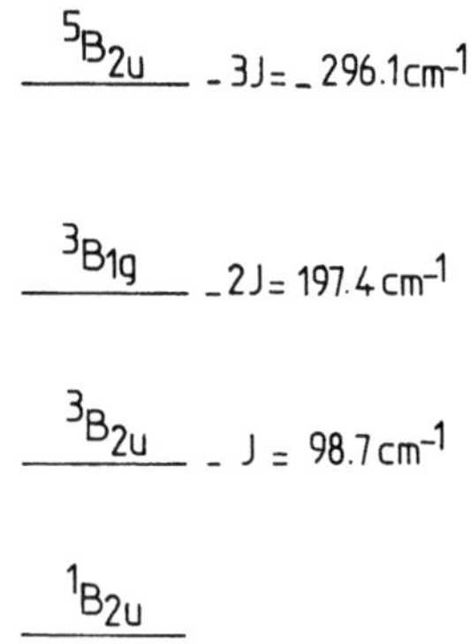

Fig. 38. Spectrum of the low lying states in $[Cu_2Ni(oxpn)_2(H_2O)_2](ClO_4)_2$ (from Ref. 131)

nitude of the phenomenon varies as dithiooxalato $C_2O_2S_2^{2-}$ > oxamido $[C_2O_2(NR)_2]^{2-}$ > oxamato $[C_2O_3(NR)]^{2-}$ > oxalato $C_2O_4^{2-}$. Let us mention a last result concerning the bis bidentate ligands $C_2X_2Y_2^{2-}$. Those bridging ligands are not at all appropriate to propagate a ferromagnetic interaction. The Cu(II)VO(II) compound of Fig. 6 illustrates this behavior. In spite of the strict orthogonality of the magnetic orbitals, the ferromagnetic stabilization is almost negligible. In the following of this article, we ought to remember that the $C_2X_2Y_2^{2-}$ bridges lead to strong antiferromagnetic interactions, provided that xy-type magnetic orbitals are available.

9.4 Cu(II)Gd(III)Cu(II) Species

Very recently, Gatteschi et al. reported on the crystal structures and the magnetic properties of several trinuclear species resulting from the condensation of two moles of tetradentate Schiff base copper(II) complexes with one mole of Gd(III) salt[134, 135]. The structure of one of these complexes, $\{[Cu(Mesalen)]_2Gd(H_2O)_3\}^{3+}$, is shown in Fig. 39. Gd(III) has a distorted pentagonal bipyramidal environment and Cu(II) a square pyramidal environment. All these Cu(II)Gd(III)Cu(II) complexes exhibit surprising and potentially very interesting magnetic properties; the $\chi_M T$ versus T plot increases upon cooling down to 1.2 K, which suggests that Cu(II) and Gd(III) interact in a ferromagnetic way. For instance, the magnetic data for the complex of Fig. 39 have been interpreted from the spin Hamiltonian:

$$\mathcal{H} = -J(\hat{S}_{Cu1} \cdot \hat{S}_{Gd} + \hat{S}_{Cu2} \cdot \hat{S}_{Gd}) - j\hat{S}_{Cu1} \cdot \hat{S}_{Cu2} \tag{89}$$

with $J = 5.5\ cm^{-1}$ and $j = -4.2\ cm^{-1}$.

If the Cu(II)Gd(III) interaction is actually ferromagnetic, the mechanism leading to this result remains for us quite obscure. Indeed, Gd(III) has the f^7 configuration, with seven magnetic orbitals. As with the d^5 configuration, the strict orthogonality of the magnetic orbitals cannot be realized (see Sect. 4.4). Moreover, the magnetic orbitals around Gd(III) are very weakly delocalized toward the bridging oxygen atoms owing to the contraction of the f orbitals. In other words, a very weak antiferromagnetic interaction would be expected by taking into consideration the same concepts as those used successfully for the 3d ions. At this stage, our feeling is that both the experimental data

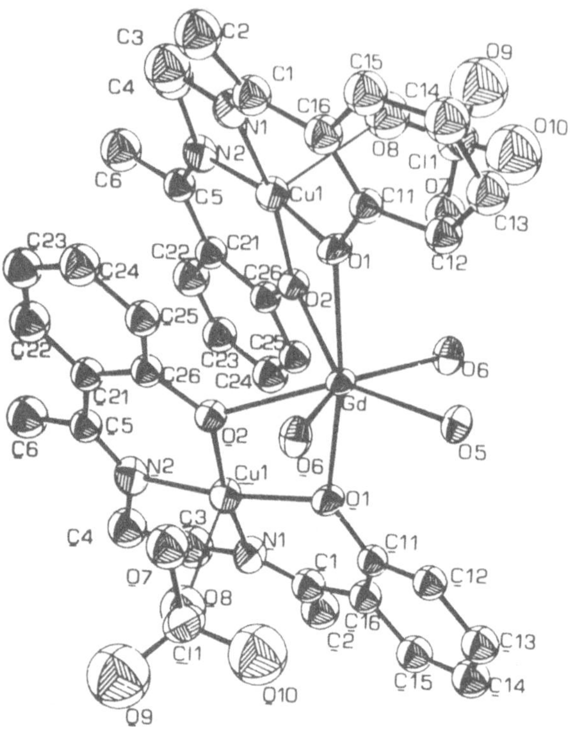

Fig. 39. Crystal structure of $\{[Cu(Mesalen)]_2Gd(H_2O)_3\}^{3+}$ (from Ref. 134)

and their interpretation have to be checked. If the ferromagnetic nature of the Cu(II)-Gd(III) interaction is confirmed, then this will prompt a search for a new mechanism of the interaction in the polymetallic systems containing rare earths, taking into account the specificity of the lanthanide ions. Perhaps, the magnetic behavior could be understood on the basis of unusually large dipolar interactions owing to the local spin 7/2 of the Gd(III) ion.

10 Regular and Irregular Spin State Structures in Trinuclear Systems

In this section, we purpose to introduce the concept of regularity and irregularity in the spin state structure. This concept already led to the synthesis of polymetallic systems in which the ground state has a very high spin multiplicity. It might be a key for the design of molecular ferromagnets, as we shall see it in Sect. 12.

10.1 Concept of Spin State Structure

Let us consider again a symmetric and linear trinuclear system ABA where the interacting centers have no orbital degeneracy and let us assume that the interaction between terminal centers is negligible with regard to the interaction between nearest neighbors. The Hamiltonian suitable for determining the spin states is:

$$\mathcal{H} = -J(\hat{S}_{A1} \cdot \hat{S}_B + \hat{S}_{A2} \cdot \hat{S}_B) \tag{90}$$

where the interactions are assumed to be purely isotropic. Several situations can be encountered concerning the distribution in energy of the spin states, according to the relative values of S_A and S_B. More precisely, three cases have to be distinguished: (i) if $2\,S_A \leqslant S_B$, we have a *regular spin state structure,* in the sense that the spin multiplicity varies monotonically with energy. If the interaction is antiferromagnetic, the ground state has the lowest spin multiplicity and the most excited state the highest spin multiplicity, the situation being reversed for a ferromagnetic interaction. The magnetic behavior arising from this regular spin state structure, it is such that upon cooling, $\chi_M T$ continuously decreases, then finally reaches a plateau corresponding to the temperature range where only the $S = S_B - 2\,S_A$ ground state is populated for $J < 0$, and continuously increases up to a plateau corresponding to the temperature range where only the $S = S_B + 2\,S_A$ ground state is populated for $J > 0$. The Cu(II)Fe(III)Cu(II) and Cu(II)Ni(II)Cu(II) complexes of Sects. 9.2 and 9.3 respectively provide two examples of antiferromagnetic interaction in trinuclear systems with regular spin state structures. As an additional example we give the spin state structure for an antiferromagnetically coupled ABA triad with $S_A = 1$ and $S_B = 5/2$ (Fig. 40). S regularly increases from $S = 1/2$ for the ground state to $S = 9/2$ for the most excited state; (ii) if $2\,S_A = S_B + 1/2$ with S_A and $S_B \neq 1/2$, the spin state structure becomes irregular and the spin multiplicity does not vary monotonically versus the energy anymore. Nevertheless, for $J < 0$, the ground state is a doublet and retains the lowest spin multiplicity; $\chi_M T$ continuously decreases upon cooling. An example of such an intermediate case is provided by the ABA triad with $S_A = 3/2$ and $S_B = 5/2$. Its spin state structure is represented in Fig. 41. For $J < 0$, the ground state is a doublet and the first excited state a quartet but the second excited state is again a doublet, etc....; (iii) if $2\,S_A > S_B + 1/2$, the spin state structure is even more irregular and for $J < 0$, the $S_g = 2\,S_A - S_B$ ground state is not that of lowest spin multiplicity. On the other hand, the most excited state is always that with the largest S ($= 2\,S_A + S_B$). This may have quite an important consequence as far as the magnetic behavior is concerned. At very high temperature, when $kT \gg |J|$, $\chi_M T$ is constant with a value corresponding to that expected for non coupled 2 A + B centers. On cooling, the

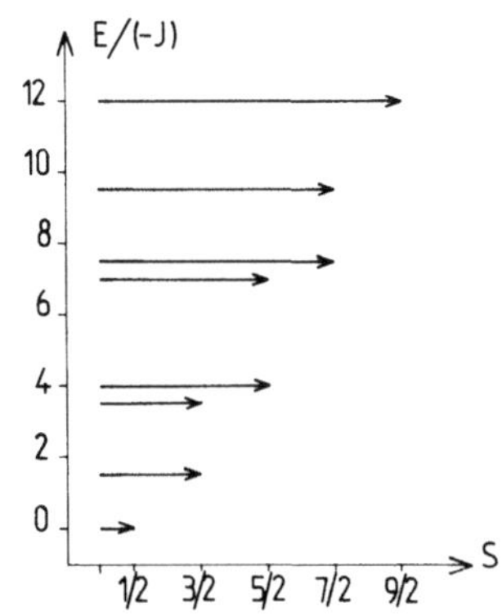

Fig. 40. Spin state structure for a symmetric ABA triad with $S_A = 1$ and $S_B = 5/2$

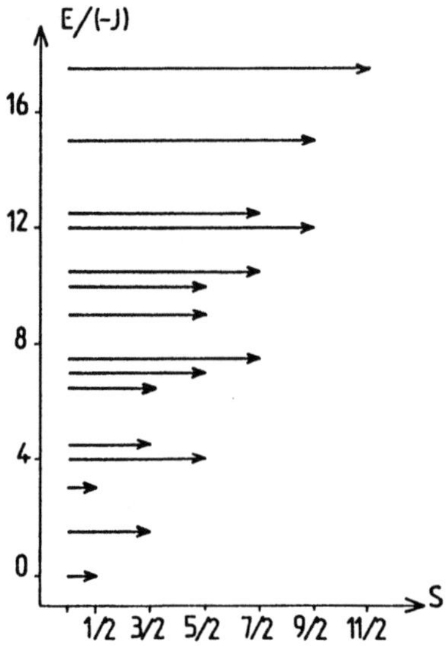

Fig. 41. Spin state structure for a symmetric ABA triad with $S_A = 3/2$ and $S_B = 5/2$

first state to be thermally depopulated is that of highest spin multiplicity, so that $\chi_M T$ decreases. In the low temperature range now, when only a few excited states are populated, further cooling depopulates states with $S > S_g$ and $\chi_M T$ may increase. Therefore, $\chi_M T$ exhibits a minimum at a non zero temperature. To our knowledge, this situation has been reported for the first time, without being discussed, by Ginsberg et al. for nickel(II) trimers[136]. It must be emphasized here that this increase of $\chi_M T$ upon cooling below the temperature of the minimum occurs although the interaction is antiferromagnetic. In this low temperature range, the magnetic behavior is reminiscent of what happens for a ferromagnetically coupled system. The larger 2 S_A with regard to S_B is, the more irregular the spin state structure. Particularly interesting in this respect is the ABA triad with $S_A = 5/2$ and $S_B = 1/2$. Its spin state structure is depicted in Fig. 42. For $J < 0$, going down in energy, S first decreases from 11/2 to 1/2 then increases from 1/2 to 9/2. It follows that a very pronounced increase of the $\chi_M T$ versus T plot may be expected in the low temperature range.

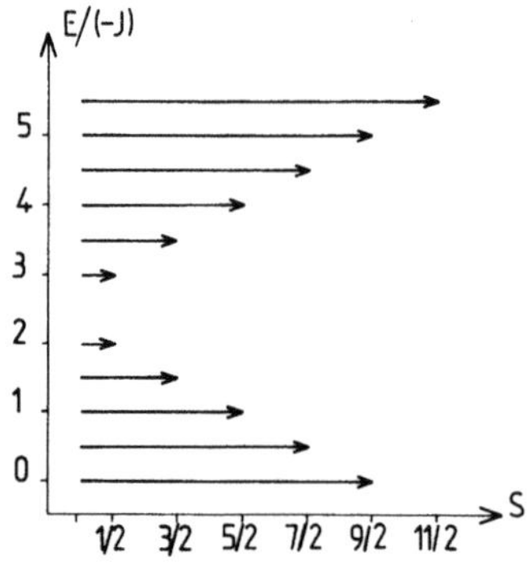

Fig. 42. Spin state structure for a symmetric ABA triad with $S_A = 5/2$ and $S_B = 1/2$

10.2 $\{[Mn(cth)]_2Cu(pba)\}(CF_3SO_3)_2 \cdot 2\ H_2O$

A complex containing the Mn(II)Cu(II)Mn(II) triad with $S_{Mn} = 5/2$ and $S_{Cu} = 1/2$ has been synthesized[137]. Its formula is $\{[Mn(cth)]_2Cu(pba)\}(CF_3SO_3)_2 \cdot 2\ H_2O$ and it is obtained by condensation of two moles of $[Mn(cth)](CF_3SO_3)_2$ with one mole of $Na_2[Cu(pba)]$. cth is d,l-5,7,7,12,14,14-hexamethyl-1,4,8,11-tetraazacyclotetradecane and pba is 1,3-propylene bis(oxamato). The skeleton of the trinuclear unit is schematized in Fig. 43. Owing to the nature of the bridges, the Mn(II)Cu(II) interaction is expected to be strongly antiferromagnetic (see Sect. 9.3), which gives rise to a 9/2 ground state with the (5/2, − 1/2, 5/2) symbolic notation for the local spins. The magnetic behavior of this compound is shown in Fig. 44. As expected, $\chi_M T$ first decreases upon cooling with a very rounded minimum about 170 K, then increases in a rather abrupt fashion up to an high plateau with $\chi_M T = 12.1\ cm^3\ mol^{-1}\ K$. This plateau exactly corresponds to what it is expected for a $S = 9/2$ isolated state ($\chi_M T = 33\ N\beta^2 g^2_{9/2(5)}/4\ k$) with $g_{9/2(5)} = 1.98$. $g_{9/2(5)}$ is related to the local tensors by:

$$g_{9/2(5)} = (6\ g_{Mn1} - g_{Cu} + 6\ g_{Mn2})/11 \tag{91}$$

Apparently the zero field splitting within the ground state is very weak since $\chi_M T$ does not exhibit any deviation from the Curie law below 10 K. This is confirmed by the EPR spectrum at 4.2 K where no fine structure is detected. This spectrum exhibits the $\Delta M_S = \pm 1$ allowed transition at $g = 1.991$ and the $\Delta M_S = \pm 2, \pm 3$ and ± 4 forbidden transitions of decreasing intensity at half, third and quarter-field respectively. The theoretical expression for the magnetic susceptibility has been established from the spin state structure of Fig. 41 and the Zeeman perturbation expressed as:

$$\mathcal{H}_Z = \beta\vec{H} \cdot (g_{Mn1} \cdot \hat{S}_{Mn1} + g_{Cu} \cdot \hat{S}_{Cu} + g_{Mn2} \cdot \hat{S}_{Mn2}) \tag{92}$$

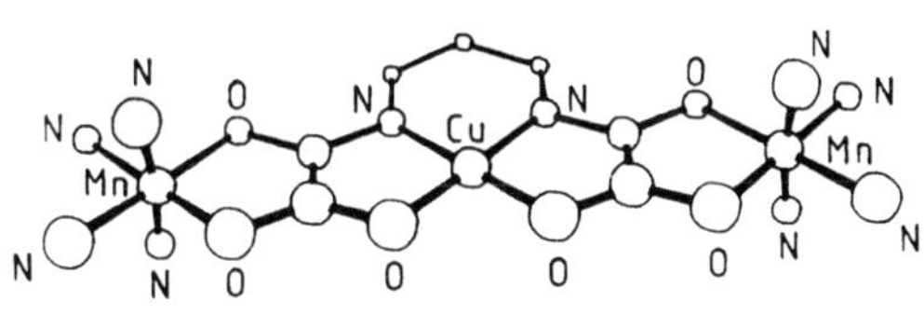

Fig. 43. Mn(II)Cu(II)Mn(II) trinuclear unit in $\{[Mn(cth)]_2Cu(pba)\}(CF_3SO_3)_2 \cdot 2\ H_2O$

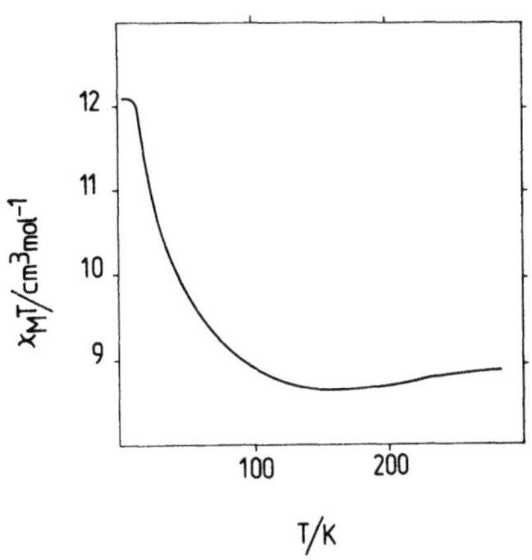

Fig. 44. $\chi_M T$ versus T plot for $\{[Mn(cth)]_2Cu(pba)\}(CF_3SO_3)_2 \cdot 2\ H_2O$

The g-factors of the eleven spin states have been calculated in terms of the local g-factors. The magnetic data are then interpreted with $J = -36.6\ cm^{-1}$, $g_{Mn} = 2.03$ and $g_{Cu} = 2.10$. This Mn(II)Cu(II)Mn(II) complex is to our knowledge the molecular unit exhibiting the largest spin multiplicity in its ground state.

A Ni(II)Cu(II)Ni(II) complex similar to that of Fig. 44 has also been synthesized, Ni(II) replacing Mn(II). It has a $S = 3/2$ ground state and the $\chi_M T$ versus T plot exhibits the characteristic minimum of this type of antiferromagnetically coupled trinuclear species with irregular spin state structure. In this compound, the isotropic interaction parameter is found equal to $J = -135\ cm^{-1}$.

11 Ordered Bimetallic Chains

We have seen in Sect. 10.1 that a trinuclear species can have an irregular spin state structure with, in case of antiferromagnetic interaction, a ground state which has not the lowest spin multiplicity. It follows that the temperature dependence of the magnetic susceptibility displays a characteristic minimum. The most spectacular example is the antiferromagnetically coupled Mn(II)Cu(II)Mn(II) triad. This concept of irregular spin state structure can be extended to more complex systems like the ordered bimetallic chains. The first compound of this kind was reported in 1981 and opened new and very exciting perspectives in the field of the magnetic systems[138]. In the forthcoming sections, we shall present first the main theoretical ideas both at a qualitative, then a quantitative level; afterwards, we shall discuss the most significant experimental results reported so far.

11.1 Theory: Qualitative Approach

To introduce the main ideas[139], we consider an ordered and regular bimetallic ring $(AB)_N$ where N may become infinitely large. A and B symbolize magnetic metal ions surrounded by their nearest neighbor ligands, including the bridging ligands, with spins S_A and $S_B \neq S_A$ respectively. We also assume that A and B have no first order angular momentum. The spin Hamiltonian in zero field appropriate to the problem is:

$$\mathcal{H} = -J \sum_{i=1}^{2N} \hat{S}_i \cdot \hat{S}_{i+1} \tag{93}$$

with:

$$S_{2i-1} = S_A$$

$$S_{2i} = S_B$$

$$\hat{S}_{2N+i} = \hat{S}_i$$

In Eq. (93), local anisotropies and anisotropic interactions have been ignored.

We suppose first that J is negative. In this case, the state of lowest energy E_g may be schematized as:

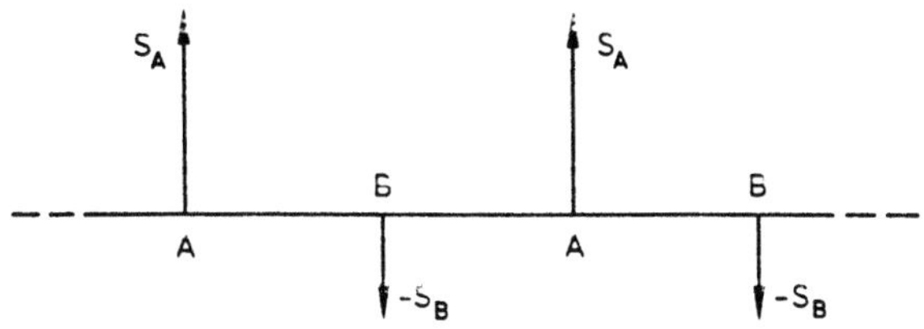

with the spin $S_g = N(|S_A - S_B|)$. The state of highest energy E_e is as:

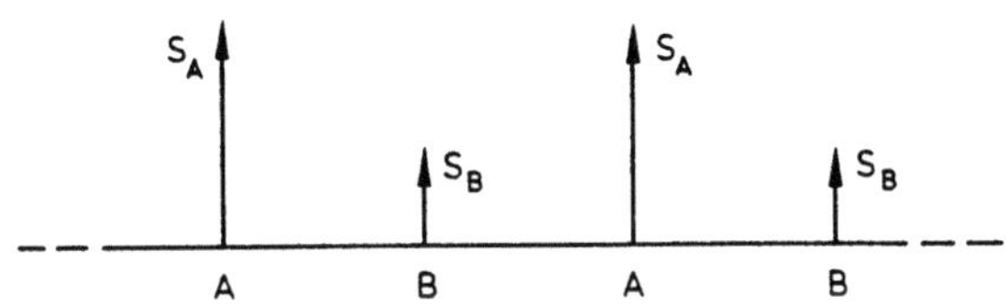

with the spin $S_e = N(S_A + S_B)$. Between the two limits E_g and E_e, states with spins less than S_g do exist. In particular, there are states with $S = 0$ if the total number of unpaired electrons $2N(S_A + S_B)$ is even; one of them is schematized hereunder:

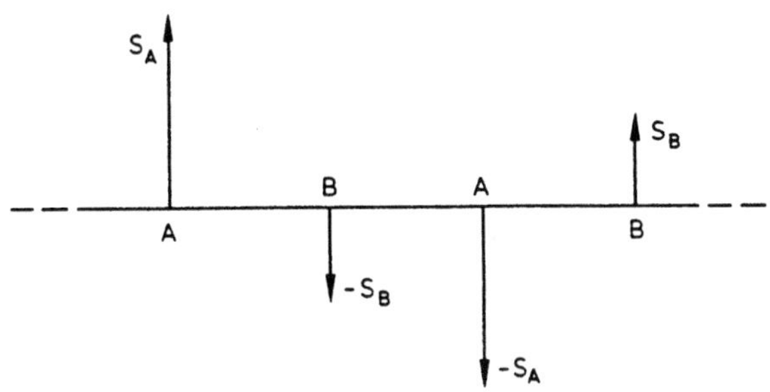

or with $S = 1/2$ if this number is odd. In other words, for any $N \neq 1$, the spin state structure is always irregular. This irregularity increases with N and $|S_A - S_B|$ since S_g is proportional to those two quantities whereas the lowest spin is always 0 or 1/2.

Let us examine now the consequences of this irregularity as far as the temperature dependence of $\chi_M T$ is concerned, χ_M being the magnetic susceptibility per AB unit. The Zeeman perturbation to add to Eq. (93) is:

$$\mathcal{H}_Z = \beta \vec{H} \cdot \sum_{i=1}^{2N} (\mathbf{g}_{A_{2i-1}} \cdot \hat{S}_{2i-1} + \mathbf{g}_{B_{2i}} \cdot \hat{S}_{2i}) \tag{94}$$

Since, in this section, we restrict ourselves to a qualitative approach, we assume that the g_A and g_B tensors are isotropic with the same principal value g. At very high temperature, when $kT \gg |J|$, $\chi_M T$ tends to the $(\chi_M T)_{HT}$ limit corresponding to the superposition of the uncoupled A and B fragments. This limit is:

$$(\chi_M T)_{HT} = \frac{N\beta^2 g^2}{3\,k} \cdot [S_A(S_A + 1) + S_B(S_B + 1)] \tag{95}$$

The low temperature limit, when $kT/|J|$ approaches zero, is:

$$(\chi_M T)_{LT} = \frac{N\beta^2 g^2}{3\,k}\,[N(S_A - S_B)^2 + |S_A - S_B|] \tag{96}$$

For $N = N_0$ defined as:

$$N_0 = \frac{S_A^2 + S_B^2 + 2\,S_<}{(S_A - S_B)^2} \tag{97}$$

where $S_<$ is the smaller of S_A and S_B, the limits $(\chi_M T)_{HT}$ and $(\chi_M T)_{LT}$ are equal. For $N > N_0$, $(\chi_M T)_{LT}$ is superior to $(\chi_M T)_{HT}$ and when N tends to the infinite, $(\chi_M T)_{LT}$ diverges. Upon cooling from the high temperatures, one depopulates first the state of highest spin multiplicity so that $\chi_M T$ decreases. Therefore one is led to the following fundamental result: *for a chain of alternated and antiferromagnetically coupled spins S_A and S_B, upon cooling, $\chi_M T$ first decreases, then reaches a minimum for a finite temperature and finally diverges when T approaches zero.* This behavior is valid for any couple of S_A and $S_B \neq S_A$, provided there is no compensation of the local magnetic moments (see Sect. 11.2). Another way to express the same result is to say that at high temperature, $\chi_M T$ tends toward the paramagnetic limit; the minimum of $\chi_M T$ corresponds to a short range order state where the spins S_A and S_B of adjacent ions are antiparallel, but without correlation between neighboring AB units. When T decreases, the correlation length within the chain increases, leading to a magnetic short range order symbolized by $(S_A - S_B)_N$. Such a situation may be defined as the *one-dimension ferrimagnetism*. Below the temperature of the minimum of $\chi_M T$, the magnetic behavior is qualitatively equivalent to what happens in a chain of N spins $|S_A - S_B|$ ferromagnetically coupled. This remark of the utmost importance will be developed further.

If the intrachain interaction is ferromagnetic ($J > 0$), the order of the spin levels is reversed and no extremum of the $\chi_M T$ versus T plot can be predicted a priori. In fact, no compound of that sort has been reported so far and in the next sections, we shall ignore this situation.

11.2 Theory: Quantitative Approach

Several quantitative approaches of the magnetic susceptibility of ordered bimetallic chains have been proposed so far. The first one consists of carrying out the calculation on $(AB)_N$ rings of increasing size with quantum spins S_A and S_B and to extrapolate to $N \rightarrow \infty$[139, 142]. This requires for each N to diagonalize the Hamiltonian (93) by using as basis sets the microstates with M_S varying from 0 or 1/2, according as $2\,N(S_A + S_B)$ is even or odd, up to $N(S_A + S_B)$, as made for the first time by Orbach[140] for a chain of spins 1/2. This leads to the energies $E_i(S)$ of the spin states and to their multiplicity. In the hypothesis where the average g-factors g_A and g_B are equal, the calculation of the magnetic susceptibility is then straightforward since all the spin states have the same g-factor and that the spin labels S are good quantum numbers, even in the presence of the Zeeman perturbation. The method is an extension of the work of Bonner and Fisher[141] concerning regular chains of spins 1/2. For $g_A \neq g_B$, not only each spin state has a given

g_S-factor but also the Zeeman perturbation couples the components of same M_S belonging to different spin states, which introduces non zero second-order Zeeman coefficients in the calculation of the susceptibility. Even after taking into account the full D_N symmetry of the problem, this method is severely limited by the storage capacity of a computer as well as by the computing time. The problem which has been the best studied is the one with $S_A = 1/2$, $S_B = 1$. Drillon et al.[142] performed the calculation up to $N = 5$ and extrapolated for $N \to \infty$. The result for $g_A = g_B$ is shown in Fig. 45. The minimum of $\chi_M T$ is obtained for $kT/|J| = 0.570$. In the very low temperature range, $\chi_M T$ diverges according to a power law $T^{-0.80}$, which coincides with the law found for a chain of ferromagnetically coupled spins 1/2. Drillon et al. also investigated the effect of $g_A/g_B \neq 1$ and found that for $g_A/g_B = 2.66$, there is an exact compensation of the magnetic moments associated with A and B respectively. The ground state becomes non magnetic and $\chi_M T$ continuously decreases down to zero upon cooling down. In a series of elegant papers, Georges, Drillon et al.[143–145] investigated in detail this compensation effect and suggested that new physical effects could be related to this compensation of the two sublattices. The $(AB)_N$ ring chain technique has also been utilized for the $S_A = 1/2$, $S_B = 5/2$[146] and $S_A = 1$, $S_B = 5/2$[147] situations but in both cases only up to $N = 3$. Unfortunately, the problem becomes untractable very quickly and alternative methods are indispensable.

The simplest method consists of considering that both $\hat{S}_A$ and $\hat{S}_B$ are classical spins with g_A^e and g_B^e effective g-factors and a J^e effective interaction parameter. The effective parameters are related to the actual parameters through:

$$g_A^e = g_A[S_A(S_A + 1)]^{1/2} \tag{98}$$

and a similar relation for g_B^e, and:

$$j^e = j[S_A(S_A + 1)S_B(S_B + 1)]^{1/2} \tag{99}$$

An analytical expression for the magnetic susceptibility has been derived by Drillon et al.[148]. This expression is:

$$\chi_M T = \frac{N\beta^2}{3\,k}\left(g^2\,\frac{1 + u}{1 - u} + \delta^2\,\frac{1 - u}{1 + u}\right) \tag{100}$$

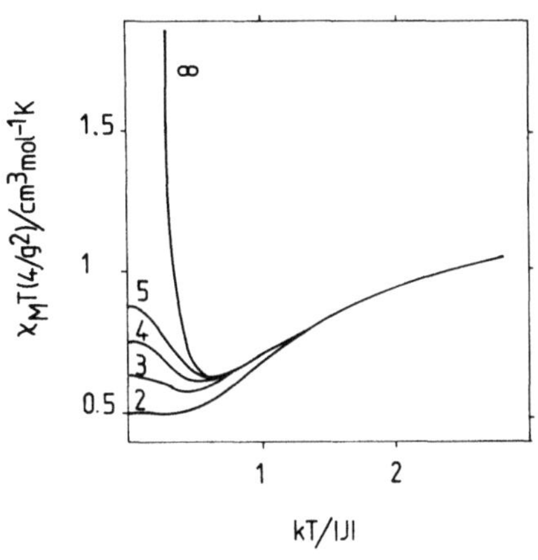

Fig. 45. $\chi_M T$ $(4/g^2)$ versus $kT/|J|$ plot for antiferromagnetically coupled $(AB)_N$ ring chains of spins $S_A = 1/2$, $S_B = 1$ (from Ref. 142)

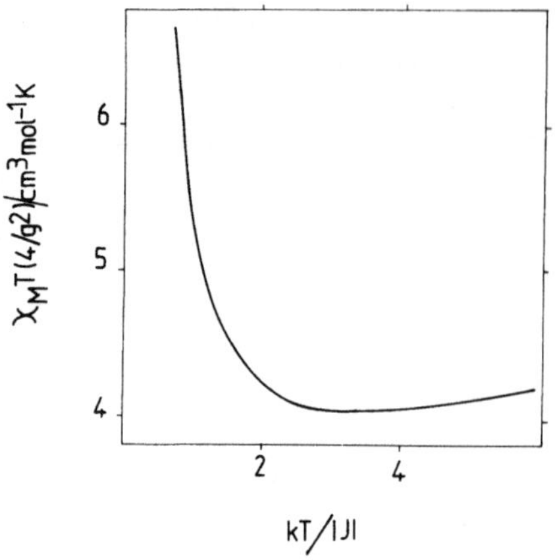

Fig. 46. $\chi_M T\ (4/g^2)$ versus kT/|J| plot for an infinite $(AB)_N$ chain with $S_A = 1/2$, $S_B = 5/2$ (from Ref. 151)

with:

$$g = 1/2(g_A^e + g_B^e)$$
$$\delta = 1/2(g_A^e - g_B^e) \tag{101}$$
$$u = \coth(j^e/kT) - (kT/j^e)$$

For $\delta = 0$, (100) reduces to the known expression for a regular chain of classical spins[149]. For $\delta \neq 0$ and $J < 0$, a minimum in the $\chi_M T$ versus T plot appears and when T approaches zero, $\chi_M T$ diverges in a ferromagnetic like fashion. The divergence law is $\delta^2 T^{-1}$. The validity of (100) is the same as the validity of the classical spin approximation. The larger S_A and S_B are, the better the relation (100). For $S_B = 5/2$, (100) is likely valid if $S_A = 2$, maybe also if $S_A = 3/2$ or 1 but it is certainly a very poor approximation when $S_A = 1/2$.

The $S_A = 1/2$, $S_B = 5/2$ case has been treated in quite an elegant way by Seiden, Verdaguer et al.[150–151]. The method consists of taking S_A as a quantum spin and S_B as a classical spin. We cannot present here the details of the calculation which is very elaborate. This calculation does not lead to an analytical expression but can be solved numerically. The following empirical relation has been proposed to fit the numerical results[152]:

$$\chi_M T = \frac{g^2}{4}(4.75 - 1.62370\,X + 2.05042\,X^2 - 4.52588\,X^3 - 8.64256\,X^4)/$$
$$(1 + 0.77968\,X - 1.56527\,X^2 - 1.57333\,X^3 - 0.11666\,X^{4.5}) \tag{102}$$

with:

$$X = |J|/kT \tag{103}$$

(102) is only valid for $J < 0$. The plot of $\chi_M T(4/g^2)$ versus kT/|J| is represented in Fig. 46. $\chi_M T$ exhibits a rounded maximum for $kT/|J| = 2.98$. Upon cooling down below the temperature of the minimum, $\chi_M T$ diverges in an abrupt way, following a law in $T^{-0.69}$. In this quantum-classical approach, the g_A and g_B factors were assumed to be equal.

In all the approaches mentioned above, the local anisotropies of the ions with S_A (or S_B) > 1/2 were neglected. Such an approximation can lead to a poor description of the low temperature range if |D/J| is not very small. In the specific case $S_A = 1/2$, $S_B = 1$, it has been shown that increasing |D/J| shifts the minimum of $\chi_M T$ toward lower temperatures[139]. An eventual anisotropic interaction would have the same effect. The local anisotropy has also been taken into account in a treatment where the interaction was assumed to be of the Ising-type, with a spin Hamiltonian in zero field as[144, 145]:

$$\mathcal{H} = -J \sum_{i=1}^{2N} \hat{S}_{z_i} \cdot \hat{S}_{z_{i+1}} \tag{104}$$

where only the components of the local spin operators along the quantum axis are involved. Such an Hamiltonian (104) may be valid for interpreting the low temperature magnetic data for an $(AB)_N$ system where both A and B are orbitally degenerate.

Finally, all the calculations deal with perfectly isolated $(AB)_N$ chains. In fact, much above the critical temperature where a three-dimension order appears, the interchain interaction may substantially modify the magnetic behavior. If the interchain interaction couples the chains in an antiparallel fashion, as it is most often the case, then the divergence of $\chi_M T$ is stopped with a maximum of $\chi_M T$ occuring just above the critical temperature. If this interchain interaction is large, the minimum in $\chi_M T$ can be hidden.

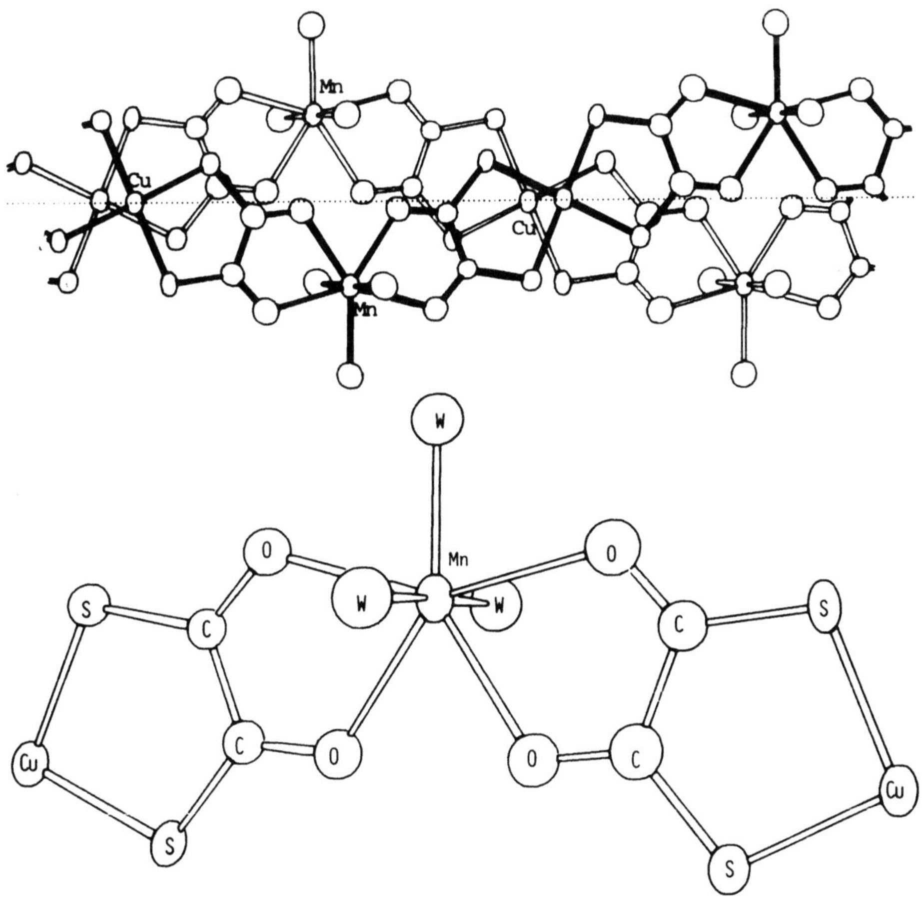

Fig. 47. Crystal structure of $CuMn(dto)_2(H_2O)_3 \cdot 4.5\ H_2O$ (from Ref. 151)

11.3 $CuMn(dto)_2(H_2O)_3 \cdot 4.5\ H_2O$

The first structurally characterized ordered bimetallic chain is $CuMn(dto)_2(H_2O)_3 \cdot 4.5\ H_2O$ with dto = dithiooxalato. This compound has been investigated by Gleizes and Verdaguer[151, 152]. The structure of two adjacent chains as well as the detail of the structure of one of the chains are shown in Fig. 47. Cu(II) is tetracoordinated with a planar environment and Mn(II) is heptacoordinated in a rather unusual fashion. The metal centers are bridged by the dithiooxalato ligand, the oxygen atoms being bound to manganese and the sulfur atoms to copper. The magnetic data (Fig. 48) closely follows the theoretical predictions. Upon cooling, $\chi_M T$ decreases, reaches a weakly pronounced minimum around 130 K and then increases as T is lowered further to 7.5 K. $\chi_M T$ is then equal to 11.3 $cm^3\ mol^{-1}$ K. Finally, below 7.5 K, $\chi_M T$ decreases rapidly. In the range 7.5–300 K, these magnetic data are well fitted by the quantum-classical approach presented in the preceding section, with $J = -30.3\ cm^{-1}$ and $g = 1.90$. The maximum of $\chi_M T$ at 7.5 K is clearly due to an antiferromagnetic interaction between ferrimagnetic like chains. Magnetization studies at 1.3 and 4.2 K show a saturation corresponding to the spin $|S_A - S_B| = 2$ per CuMn unit. The relatively large $|J|$ value found in $CuMn(dto)_2(H_2O)_3 \cdot 4.5\ H_2O$ is due to the remarkable efficiency of the dithiooxalato bridge to propagate an antiferromagnetic interaction between metal centers with xy-type magnetic orbitals.

11.4 $CuMn(pba)(H_2O)_3 \cdot 2\ H_2O$ and Related Compounds

Another example of Cu(II)Mn(II) ordered bimetallic chain is provided by $CuMn(pba)(H_2O)_3 \cdot 2\ H_2O$, with pba = 1,3-propylene bis (oxamato). The structure of the chain is given in Fig. 49 and that of the crystal lattice in Fig. 50[153]. Cu(II) is here in a square pyramidal environment and Mn(II) in a more classical octahedral environment. The metal centers are aligned along the b axis of the orthorhombic structure, whereas in the previous example they formed zig zag chains. Within the chains, two nearest neighbor metal ions are bridged by an oxamato group with a Cu ... Mn separation of 5.41 Å. The chains are linked together through hydrogen bonding involving the oxygen atoms of

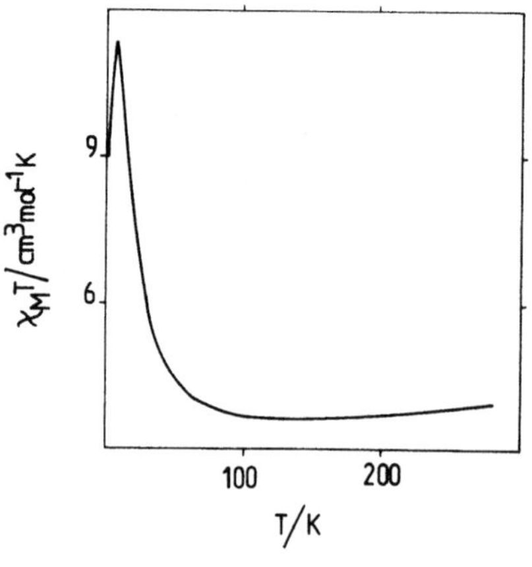

Fig. 48. $\chi_M T$ versus T plot for $CuMn(dto)_2(H_2O)_3 \cdot 4.5\ H_2O$ (from Ref. 151)

the bridge and coordinated as well as the non coordinated water molecules. The relative positions of two adjacent chains are such that the shortest interchain metal... metal separations are Cu... Cu and Mn... Mn.

As expected, the $\chi_M T$ versus T plot exhibits a rounded minimum about 115 K and below this temperature a very fast increase upon cooling. Again a maximum of $\chi_M T$ is

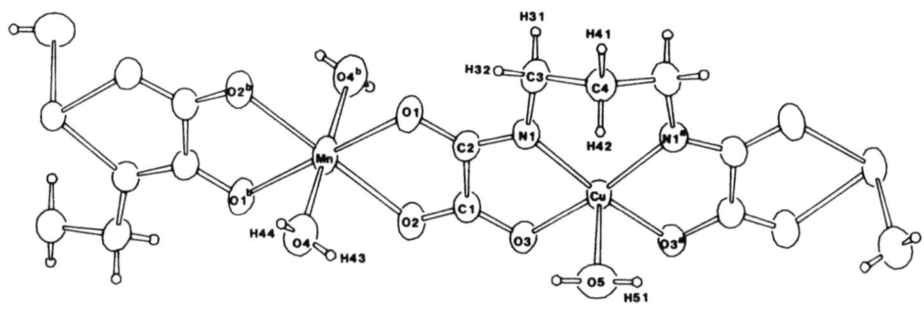

Fig. 49. Crystal structure of $CuMn(pba)(H_2O)_3 \cdot 2\ H_2O$ (from Ref. 153)

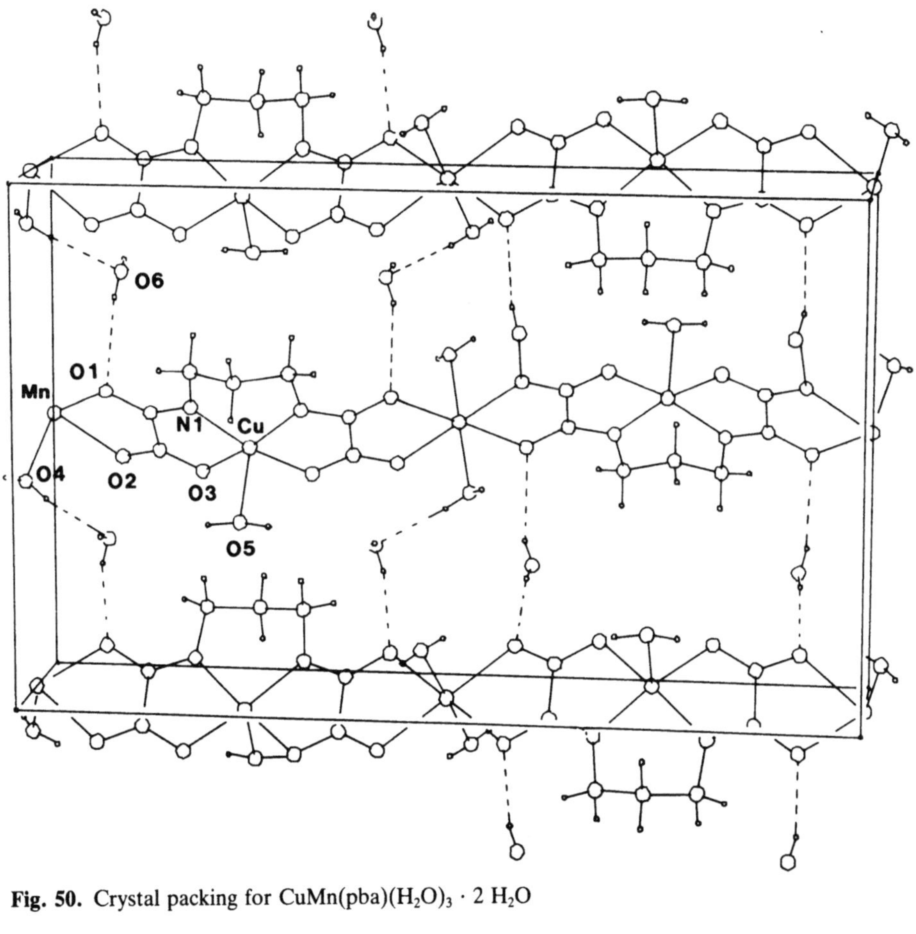

Fig. 50. Crystal packing for $CuMn(pba)(H_2O)_3 \cdot 2\ H_2O$

observed, but at a much lower temperature than in the previous case, namely 2.3 K instead of 7.5 K. This indicates that the interchain interaction is weaker in this compound than in the previous one. The fitting of the magnetic data above 4.2 K leads to J = $-23.4\ cm^{-1}$ with g = 1.97[154]. This J value, although substantial, is significantly smaller in absolute value than that found in $CuMn(dto)_2(H_2O)_3 \cdot 4.5\ H_2O$. This is because the magnetic orbital around copper(II) is more delocalized toward the sulfur atoms of the dithiooxalato bridge in the dto derivative than toward the nitrogen and oxygen atoms of the oxamato bridge in the pba derivative, due to the weak electronegativity of sulfur relative to nitrogen and oxygen.

Let us examine now the mechanism of the interchain coupling. Owing to the relative positions of two adjacent chains, the dominant interchain interactions are Cu(II)Cu(II) and Mn(II)Mn(II). If these interactions are antiferromagnetic, then the spin structure will be:

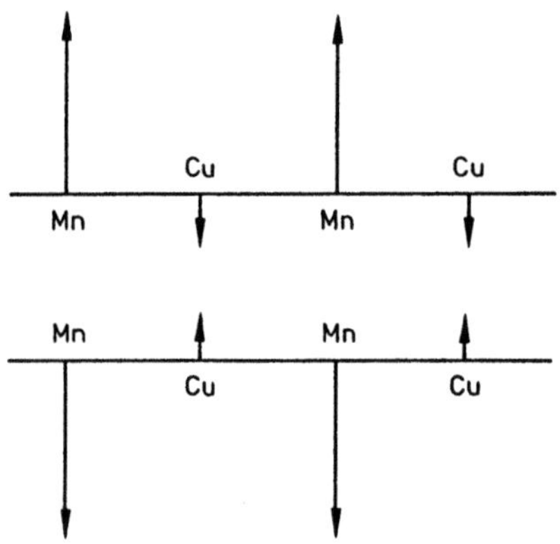

with a cancellation of the spins at the scale of the lattice. This leads to a decrease of $\chi_M T$ below 2.3 K. Now, if by using some chemical stratagem we can displace every other chain in the crystal lattice by half of a repeat unit along the b axis, then the interchain interactions will occur between metal ions of different nature and, provided that these interactions are again antiferromagnetic, the spin structure will be:

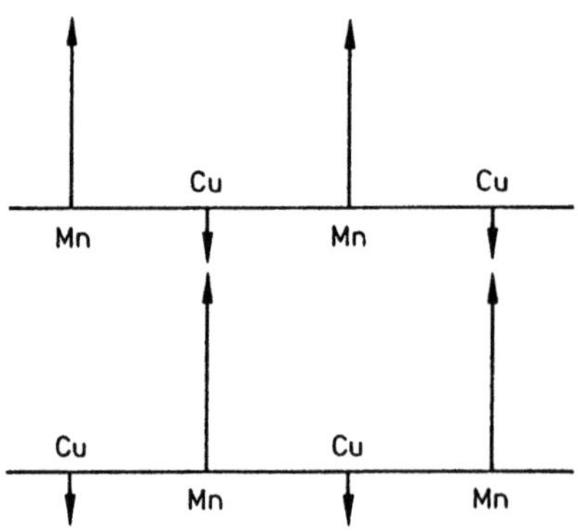

with, at the scale of the lattice, all the 5/2 spins aligned along the same direction. Such a system should exhibit a three-dimension ferromagnetic (or ferrimagnetic) order below a critical temperature. Such a situation could be at the origin of the quite remarkable magnetic properties of CuMn(pbaOH) $\cdot 3\ H_2O$[155]. Chemically, the only difference between this compound and $CuMn(pba)(H_2O)_3 \cdot 2\ H_2O$ is the replacement of the central methylene group of the propylene chain by CHOH. Above 60 K, the pba and pbaOH derivatives exhibit the same magnetic properties with the characteristic minimum of $\chi_M T$ around 115 K. On the other hand, below 60 K, $\chi_M T$ for CuMn(pbaOH) $\cdot$ 3 H_2O increases in a much more abrupt fashion and reaches an extraordinary high value at 4.2 K. $\chi_M T$ is then larger than 100 $cm^3\ mol^{-1}$ K! Magnetization studies in very low field (H = 0.03 G) show that the compound orders ferromagnetically at 4.6 K. As it is usual, the ferromagnetic order is distroyed by the magnetic field, even when H is as small as 1 G. CuMn(pbaOH) $\cdot$ 3 H_2O may be considered as one of the first genuine molecular ferromagnets. Its crystal structure has now been solved. It actually shows a displacement of every other chain by roughly half of a repeat unit along one of the directions perpendicular to the chain axis.

The $CuNi(pba)(H_2O)_3 \cdot 2\ H_2O$ chain with Ni(II) replacing Mn(II) in the structure of Fig. 49 has also been prepared and studied[154]. The $\chi_M T$ versus T plot is shown in Fig. 51. It exhibits the characteristic minimum at 83 K and below this temperature $\chi_M T$ increases rapidly up to a maximum at 7.3 K, reflecting again an interchain interaction. The fitting of the magnetic data between 10–300 K using the ring chain approach leads to J = $-82.7\ cm^{-1}$ with g = 2.17.

Another Cu(II)Ni(II) chain, of formula $CuNi(ox)_2 \cdot 4\ H_2O$ with ox = oxalato, has also been described[139]. The available structural information does not come from X-ray diffraction but from an EXAFS study which is consistent with ordered Cu(II)Ni(II) chains and oxalato bridges between the metal centers. The chains, however, are poorly isolated from each others. As a matter of fact, each copper atom of a chain *sees* two other copper atoms at 4 Å, belonging to other chains. The poor one-dimension character likely explains why the $\chi_M T$ versus T plot does not display the expected minimum. In fact $\chi_M T$ continuously decreases upon cooling. Since the low temperature data are severely affected by the relatively strong interchain interaction, the intrachain interaction parameter J has been deduced from the magnetic data above 50 K and found equal to $-52.7\ cm^{-1}$. The antiferromagnetic interaction is a bit less pronounced than in CuNi(pba)$(H_2O)_3 \cdot 2\ H_2O$. This well corresponds to the order oxamato > oxalato given in Sect. 9.3.

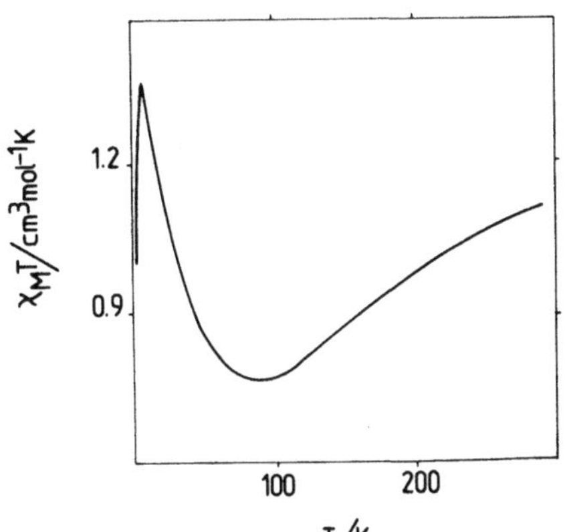

Fig. 51. $\chi_M T$ versus T plot for $CuNi(pba)(H_2O)_3 \cdot 2\ H_2O$

11.5 AB(edta)(H$_2$O)$_4$ · 2 H$_2$O

A completely different family of ordered bimetallic chains has been investigated by Drillon et al. The general formula is AB(edta)$(H_2O)_4$ · 2 H_2O where edta is ethylenediaminetetracetato and A and B two divalent ions. All the compounds are reported to be isomorphous with A_2(edta)$(H_2O)_4$ · 2 H_2O, A = Zn[156] and Ni[157], of which the crystal structures have been solved. The structure of the nickel(II) compound is schematized in Fig. 52. There are two distinct coordination sites (Fig. 52). In the first, a metal ion, say A, is in a distorded octahedral environment involving the oxygen atoms of the four carbonyl groups and the nitrogen atoms of edta. In the second, B is in a less distorted octahedron formed by four water molecules and two oxygen atoms of adjacent A(edta) units. Thus, two of the four carboxylic groups of edta are bidentate bridges. All together, zig zag chains [A(edta) – B$(H_2O)_4$]$_n$ are formed. The two coordination sites have some selectively. For instance, Cu(II) and Ni(II) prefer the chelated sites A and when these sites are fully occupied, Co(II) or Mn(II) goes into the hydrated sites B.

The most accurately studied compound of this family is the Ni(II)Mn(II) one[147]. The χ_MT versus T plot is shown in Fig. 53. It exhibits the expected minimum, but at a very low

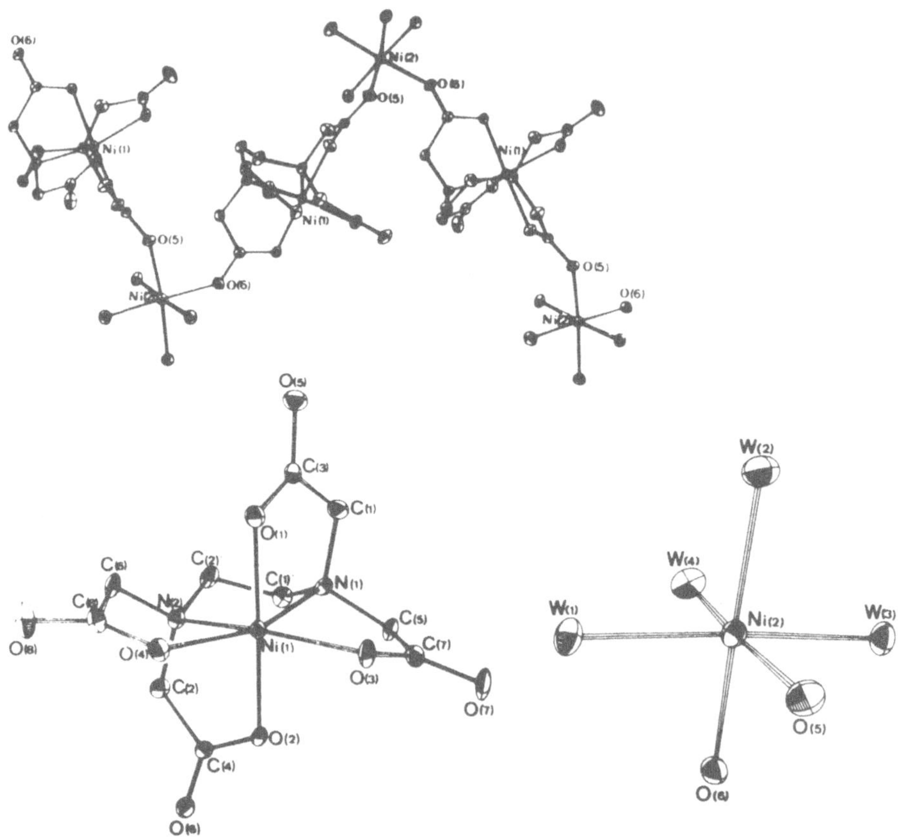

Fig. 52. Crystal structure of Ni_2(edta)$(H_2O)_4$ · 2 H_2O (from Ref. 157)

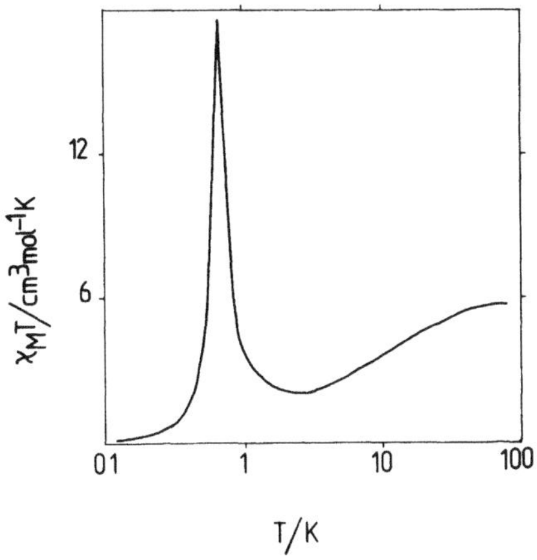

Fig. 53. $\chi_M T$ versus T plot for $NiMn(edta)(H_2O)_4 \cdot 2\ H_2O$. The scale is semilogarithmic (from Ref. 147)

temperature, namely 2.5 K, then a rapid increase upon cooling below 2.5 K and a sharp maximum at 0.665 K. At lower temperatures, $\chi_M T$ tends to zero when T approaches zero. The low temperature of the minimum of $\chi_M T$ indicates that the intrachain interaction is much weaker than in the compounds described in Sects. 11.3 and 11.4, which is not surprising owing to the structure of the $AB(edta)(H_2O)_4 \cdot 2\ H_2O$ chains. The maximum of $\chi_M T$ at 0.665 K is due to the onset of a three-dimension antiferromagnetic order. Heat capacity measurements confirm that a long range magnetic order occurs at this temperature. Indeed, the heat capacity curve exhibits a characteristic λ-peak at 0.660 K. The magnetic data have been fitted with both the classical spin approach (relations (98)–(101)) and the ring chain approach. The results of this latter model are: $J = -1.04\ cm^{-1}$, $g_{Ni} = 2.39$ and $g_{Mn} = 1.95$.

The magnetic data for $CoMn(edta)(H_2O)_4 \cdot 2\ H_2O$ have also been recorded[147]. The characteristic minimum of $\chi_M T$ is found around 4 K and the onset of the antiferromagnetic order is observed at 1.06 K through heat capacity measurements. In the low temperature range, say below 20 K, only the local Kramers doublet of Co(II) may be thermally populated. Therefore, in this temperature range, it is justified to treat Co(II) as an ion with an 1/2 local effective spin. To interpret the magnetic data, an Ising exchange model was used (see relation (104)) with a quantum spin 1/2 for Co(II) and a classical spin for Mn(II). J was then found equal to $-1.33\ cm^{-1}$.

Magnetic studies for the Ni(II)Co(II) and Cu(II)Co(II) compounds have been reported[158], as well as the theoretical treatments suitable for those cases.

12 Systems with Subtle Spin Orders

We hope that the reader is now convinced of the interest of the ordered bimetallic chains. Their existence in itself demonstrates the feasibility of preparing systems with subtle spin orders, in spite of apparent kinetic and entropic hindrances[138]. This concept of subtle spin order might lead to new molecular materials exhibiting important physical properties.

One of the main challenges in the field of the molecular materials is the design of molecular ferromagnets. For that, a strategy immediately comes in mind, namely the achievement of ferromagnetic interactions between the nearest neighbor ions at the scale of the whole lattice. If so, the system will order ferromagnetically below a critical temperature. Such ferromagnetic interactions between nearest neighbors may arise either from the strict or from the accidental orthogonality of the magnetic orbitals. The strict orthogonality is apparently a very elegant approach but it works with only a few couples of metal ions and requires relatively high site symmetries. In principle, the most favorable couple would be chromium(III)-nickel(II), both ions in octahedral surroundings. Cr(III) has three unpaired electrons occupying the t_{2g} orbitals and Ni(II) two unpaired electrons occupying the e_g orbitals. All the $\langle t_{2g}|e_g\rangle$ overlap integrals are zero and the Cr(III)Ni(II) interaction is expected to be ferromagnetic. Unfortunately, so far, no Cr(III)Ni(II) molecular system or extended lattice has been synthesized. As for the accidental orthogonality, it depends in quite a critical manner on some structural parameters and it is certainly not possible to predict a priori what kind of chemical process will lead to such an accidental orthogonality. At best, one knows that with some geometries, the accidental orthogonality may eventually occur. In a certain sense, the achievement of a ferromagnetic interaction between two nearest neighbor ions requires going against a natural trend which favors the pairing of the electrons in molecular orbitals of low energy. Therefore, it was important to find an alternative strategy for designing molecular systems with an high spin multiplicity in the ground state and to develop this strategy up to the molecular ferromagnets.

Such a strategy emerges from the concept of irregular spin state structure. It consists of imposing the parallel alignment of local spins 5/2 (Mn(II) or Fe(III)) owing to an antiferromagnetic interaction with local spins 1/2 (Cu(II)). In some way, the small spins (1/2) polarize the big spins (5/2) along a common direction. This leads to a 9/2 ground state for a Mn(II)Cu(II)Mn(II) trinuclear unit, to a one-dimension ferrimagnetic short-range order for the Cu(II)Mn(II) chain and to the onset of a ferromagnetic order when a small Cu(II)Mn(II) interchain interaction is superimposed on the dominant Cu(II)-Mn(II) intrachain interaction. It is quite remarkable that the $S_A = 1/2$, $S_B = 5/2$ couple is the most appropriate to obtain a ferromagnetic like behavior in the frame of this strategy whereas the interaction between those two ions is most likely antiferromagnetic (see Sect. 4.4). In fact, the more pronounced the Cu(II)Mn(II) antiferromagnetic interaction, the more efficient the polarization of the 5/2 spins along the common direction. In this respect, the bis bidentate ligands like dithiooxalato or oxamato are particularly suitable to this purpose. Not only do they give strong antiferromagnetic interactions (see Sect. 9.3), but also their chemistry is very flexible.

To show all the potentialities of this Cu(II)Mn(II) strategy, we would like to stress a last point. Let us imagine that we want to design the trinuclear species with 3 d ions, of which the ground state has the highest spin multiplicity. First, we seek strict orthogonality of the magnetic orbitals. The best situation is then Cr(III)Ni(II)Cr(III) with the three ions in octahedral surroundings. The ground state will have spin $S = 4$. In spite of the two Cr(III)Ni(II) interactions, the result is inferior to that obtained with Mn(II)Cu(II)Mn(II), of which the ground state has the spin $S = 9/2$. This strategy consisting of polarizing the big spins is not a makeshift but really seems to be the most efficient in the perspective of molecular ferromagnets. It is the most efficient also because it leans on the fact that the nature prefers the antiferromagnetic interaction – it is a border line form of chemical bond – instead of opposing this fact.

To conclude this section, we want to point out that spin orders more subtle than the alternation $(S_A, S_B)_N$ can be obtained. For instance, the compound following has been synthesized:

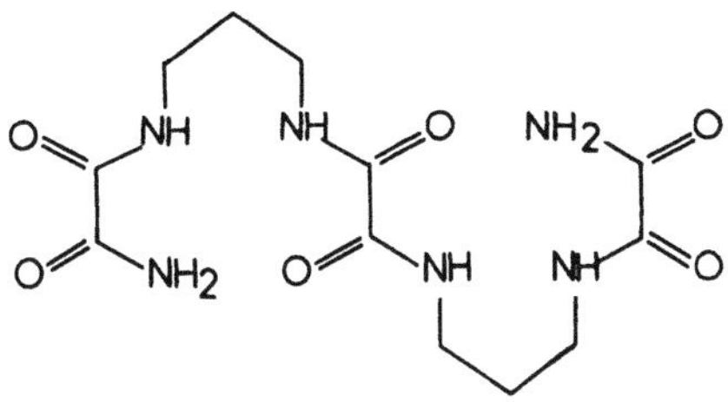

It reacts with two Cu(II) in presence of six OH^- to give after hydrolysis of the terminal primary amine groups:

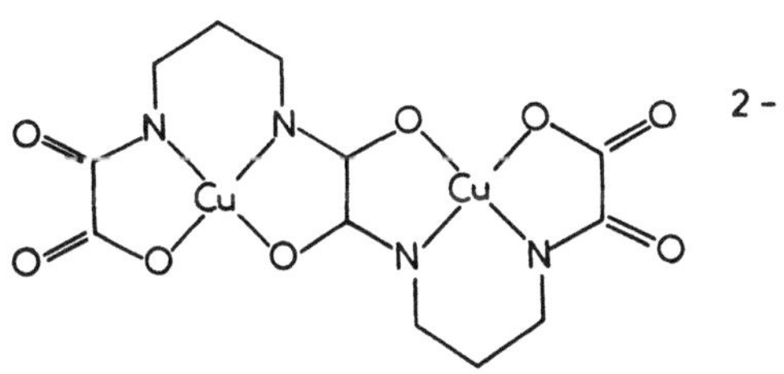

from which the $MnCu_2(bapo)(H_2O)_4 \cdot 2\,H_2O$ chain schematized below can be synthesized[153] (bapo stands for N,N′-bis(oxamato-1,3-propylene)oxamide).

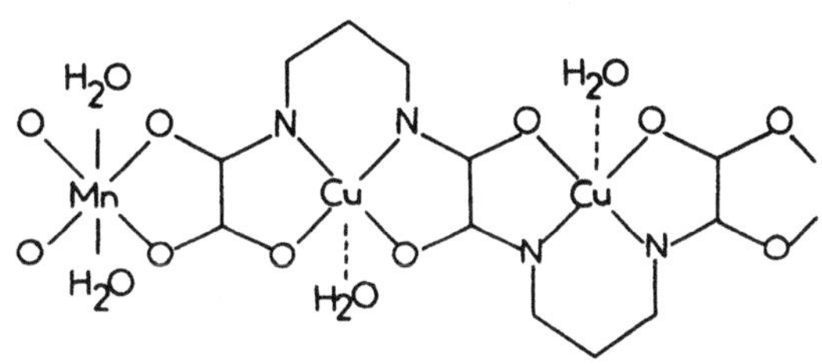

All the interactions between nearest neighbor ions are antiferromagnetic. Hence, the spin structure in the ground state is:

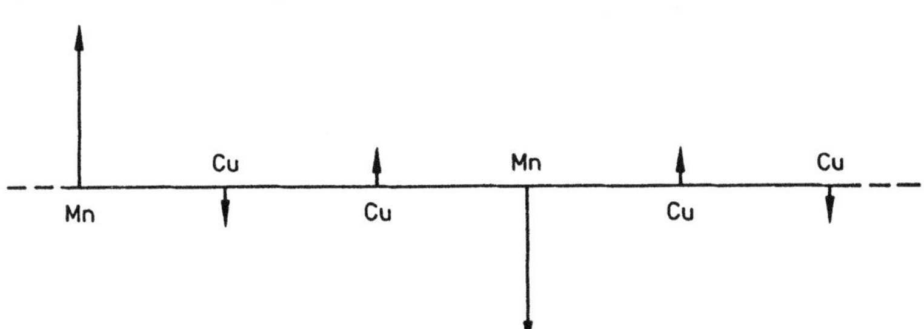

This state is non magnetic and $\chi_M T$ (per Cu_2Mn unit) is expected to decrease continuously upon cooling and to tend to zero as T approaches zero, which is confirmed by the

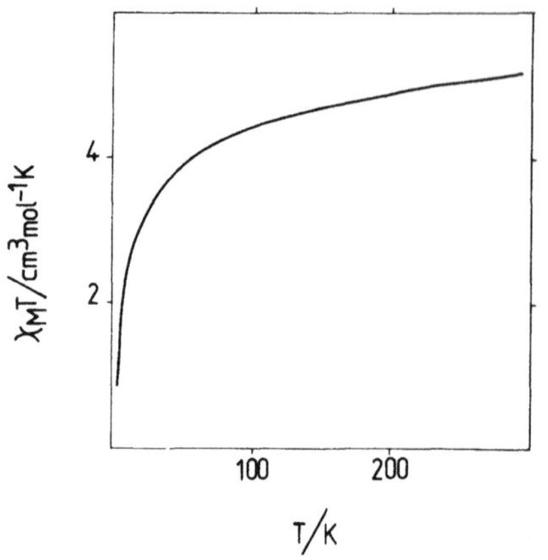

Fig. 54. $\chi_M T$ versus T plot for $MnCu_2(bapo)(H_2O)_4 \cdot 2\ H_2O$ (from Ref. 153)

experimental data (Fig. 54). Seiden[159] is now expanding his quantum-classical approach valid for the Cu(II)Mn(II) chains in order to treat quantitatively systems of this kind.

Up to recently, in the field of the low dimension compounds, the most peculiar and interesting physical properties had been predicted by the theoretical physicists before the synthesis of the compounds. This is in particular the case for the Peierls instability in the 1 − D conducting systems and for the spin-Peierls dimerization in the 1 − D magnetic systems[160]. It seems that now the situation is somewhat reversed. The chemists synthesize systems which the physicists had never thought of before, so the quantitative theory comes after. We believe that this process will be more and more frequent. The possibilities of the chemist in terms of new types of compounds are almost without limit, if not his own imagination. In the field of the magnetic systems with subtle spin orders, very spectacular developments may be expected.

13 Conclusion and Outlook

To start with this last section, we want to underline a point concerning the status of the field of the heteropolymetallic systems. This field cannot be considered as a simple extension of that of the homopolymetallic systems. If it is true that the study of the heteropolymetallic compounds utilizes all the concepts valid for the systems with only one kind of magnetic ions, it is also true that this new field allowed one to introduce several important and novel concepts. Let us briefly recall some of them: (i) the strict orthogonality of the magnetic orbitals. Such an orthogonality could in principle occur in homobinuclear complexes. This would be, for instance, the case for a copper(II) complex with two planar chromophores bridged by a tetrahedral ligand:

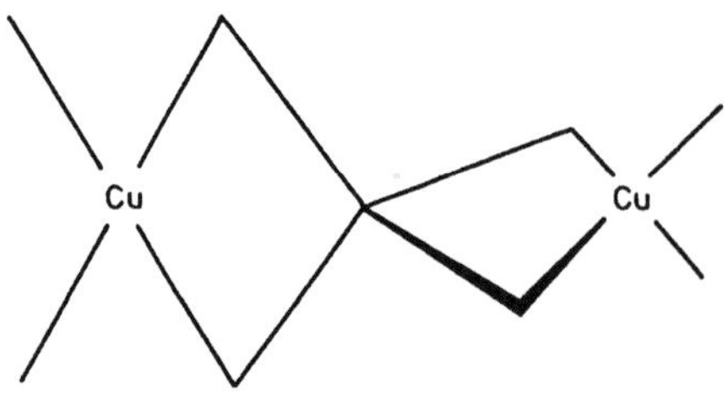

To our knowledge, no complex of that sort has been reported; (ii) the overlap density between orthogonal magnetic orbitals. If the role of the two-electron exchange integral in the ferromagnetic stabilization is known, the relation between the magnitude of this integral and the overlap density map emerges from the study of the Cu(II)VO(II) complexes; (iii) the relations between molecular and local **g, D** and **A** tensors. These relations are not specific to the systems with several kinds of metal ions but those systems are well adapted to test their range of validity. So, to establish relations (20)–(26), it is assumed that not only the isotropic interaction is much larger than the other perturbations, but also that the pair functions can be constructed as products of local functions. This latter condition corresponds to the Heitler-London approach for the weakly coupled systems. When the isotropic interaction becomes very large, the Heitler-London approach may be not valid anymore. In other words, the validity of the relations (20)–(26) vanishes when J is too small with regard to the other perturbations (particularly the local anisotropies) on the one hand, and when J is too large on the other. It is worthy of note that the range of validity of (20)–(26) is the same as that of the orbital model presented in Sect. 2.3; (iv) the concept of irregular spin state structure, leading to molecular systems with a high spin multiplicity in the ground state in spite of antiferromagnetic interactions between nearest neighbor ions. The one-dimension ferrimagnetism is an obvious extension of this concept. The experimental signature of an irregular spin state structure, in a molecular unit like a trinuclear species or a one-dimension system, is the minimum of the $\chi_M T$ versus T plot at a finite temperature; (v) the subtle spin order is both a concept and a dream. The concept is that to a new spin order are associated new magnetic properties. The dream is to play with the spins in the design of magnetic systems in the same way that chemists for a long time have known to play with the atoms in the synthesis of molecules.

The field of the heteropolymetallic systems covers the synthesis, the investigation of the physical properties and the theoretical approach. The synthesis of these systems is still a challenge. This is particularly true for the systems with more than three metal ions, including the one-dimension systems. The only tetranuclear species described in this article (Sect. 6.3) results from a peculiar packing of two binuclear units, that was obviously not expected. In a more general way, the key to design systems with a high nuclearity or with a subtle spin order is the synthesis of bricks containing one or a few metal ions that could be assembled in a perfectly controlled manner to build the expected edifice.

The determination of the energy and the nature of the low lying states apparently does not require any new technique with regard to those used for the homopolymetallic systems with the most extensively used being magnetic susceptibility. In this respect, we want to point out again how important it is to have accurate data. When studying the magnetic properties of a system with a singlet ground state like a copper(II) dimer, the χ_M versus T plot exhibits a characteristic maximum and an uncertainty in the actual temperature of this maximum roughly leads to the same uncertainty in J, since J is proportional to this temperature. On the other hand, when studying the magnetic behavior of a system of which the ground state is magnetic (this is the case of most of the heteropolymetallic systems) the χ_M versus T plot varies continuously whatever the sign of J may be. The important plot is actually $\chi_M T$ versus T. A small uncertainty in T very much affects the shape of this plot, particularly in the low temperature range, and the uncertainty in J deduced from the magnetic data is much larger than the uncertainty in T. In this article, we have only depicted average magnetic susceptibility data and actually we we do not know any report of magnetic anisotropy measurements on heteropolymetallic compounds. Such measurements could provide important new informations in the following two situations: (i) when at least one of the interacting ions has first order orbital degeneracy. In fact, in such a case (Sect. 8.1), the magnetic anisotropy data are almost indispensable for analysing the spectrum of the pair states; (ii) when the ground state is split in zero field. Accurate average magnetic data can often be fitted in the same satisfying way with both a positive and a negative value for the zero field splitting parameter. Anisotropy data unambiguously specifies the sign of the phenomenon. We have seen how the EPR technique is complementary to the magnetism so that the two techniques should be considered as indissociable for this field. Till now, Inelastic Neutron Scattering, a very elegant technique giving directly the energy gaps between low lying states, has not been used for heteropolymetallic systems[161)]. This is regrettable because this technique would be useful here, particularly when there are more than two thermally populated states. Several attempts have been performed to detect the singlet-triplet energy gap in copper(II) dimers through Raman Spectroscopy[162)]. None has been fully conclusive. This is not surprising since the transition is spin (and often parity) forbidden. When the two halves of the dimer are symmetry related, there is no mechanism for mixing singlet and triplet states. The situation could be much more favorable in an heterobinuclear complex. In a Cu(II)Ni(II) system, for instance, the doublet and quartet low lying states couple through the local anisotropy of Ni(II). When $D_{Ni}/|J|$ is not to small, the transition between these states could perhaps be detectable.

As far as the theoretical approach is concerned, three levels have to be considered: (i) the phenomenological level providing the theoretical expressions for the magnetic susceptibilities. Only the case of the one-dimension compounds presents real difficulties; (ii) the qualitative approach of the nature and the order of magnitude of the isotropic interaction. In this respect, the orbital model based on the concepts of natural magnetic orbitals and overlap densities between pairs of magnetic orbitals has borne witness to its efficiency[132)]. We think that this model provides simple and useful tools for experimentalists looking to synthesizing new systems with predictable magnetic properties. If so, it serves its purpose; (iii) the quantitative approach of J through ab initio calculation techniques. Whereas significative progresses have been achieved for the copper(II) binuclear complexes along this line[163–165)], all remains to do for the heteropolymetallic compounds. We already underlined in Sect. 2.2 that no method has yet been proposed to obtain the orthogonalized magnetic orbitals of an heteropolymetallic system from the molecular orbitals, or the Vannier functions from the Bloch functions if referring to one-dimension systems.

At this latest stage of our work, it is worthwhile to insist upon the challenge represented by the design of molecular ferromagnets, both for organic[166, 167)] and inorganic chemists. MnCu(pbaOH) · 3 H_2O is apparently the first compound of this type, ordering ferromagnetically at 4.6 K. This result is quite preliminary in the sense where neither the crystal structure of the compound, nor its other physical properties are yet known. The thorough investigation of this compound as well as the design of other systems of this kind with, if possible, higher magnetic order temperature, are two directions along which several groups are engaged. The difficulty along this way arises from the fact that there is no magnetic order at one or two dimensions, but only at three dimensions. Therefore, the chemist has to work in terms of crystal engineering, and not only of molecular engineering.

The field of the heteropolymetallic magnetic systems is still rather new. As a matter of fact, the first EPR spectrum of an heteropair appeared in 1967 and the first preliminary magnetic data dealing with compounds of this type in 1968. This article is, to our knowledge, the first review on the subject. It largely concerns heterobinuclear complexes and covers the field up to the first attempts to obtain molecular ferromagnets. One may anticipate that, in a few years, another review will be necessary and that its main emphasis will be the systems of higher nuclearity and the extended systems with subtle spin orders.

Acknowledgements. A part of the work described in this article has been carried out in Orsay by my coworkers and colleagues of the Laboratory. Their names appear in the references listed. To all of them, I want to express may deepest gratitude. I would also like to thank all the other colleagues who collaborated with me on this subject. A special mention ought to be done to D. Gatteschi and the whole group of Florence. The collaboration with them has been particularly successful and rewarding. In other respects, their specific contribution to the field is of the utmost importance. Finally, it is a great pleasure for me to specify that this article has been written during a sabbatical leave in the State University at Leiden. The friendly and stimulating atmosphere of J. Reedijk's group has been for me an invaluable help. To J. Reedijk and all the colleagues of the group, I want also to express my thanks.

14 References

1. Casellato, U., Vigato, P. A.: Coord. Chem. Rev. *23,* 31 (1977)
2. Groh, E.: Israel J. Chem. *15,* 277 (1976/1977)
3. Lintvedt, R. L., Glick, M. D., Tomlonovic, B. K., Gavel, D. P.: Inorg. Chem. *15,* 1646 (1976)
4. Glick, M. D., Lintvedt, R. L., Gavel, D. P., Tomlonovic, B.: ibid. *15,* 1654 (1976)
5. Lintvedt, R. L., Ahmad, N.: ibid. *21,* 2356 (1982)
6. Coucouvanis, D.: J. Am. Chem. Soc. *92,* 707 (1970)
7. Coucouvanis, D., Baenziger, N. C., Johnson, S. M.: ibid. *95,* 3875 (1973)
8. Coucouvanis, D., Baenziger, N. C., Johnson, S. M.: Inorg. Chem. *13,* 1191 (1974)
9. Coucouvanis, D., Piltingsrud, D.: J. Am. Chem. Soc. *95,* 5556 (1973)
10. Hollander, F. J., Coucouvanis, D.: Inorg. Chem. *13,* 2381 (1974)
11. Caffery, M. L., Coucouvanis, D.: J. Inorg. Nucl. Chem. *37,* 2081 (1975)
12. O'Bryan, N. B., Maier, T. O., Paul, I. C., Drago, R. S.: J. Am. Chem. Soc. *95,* 6640 (1973)
13. Leslie, K. A., Drago, R. S., Stucky, G. D., Kitko, D. J., Breese, J. A.: Inorg. Chem. *18,* 1885 (1979)
14. Moriya, T. in: Magnetism (Rado, G. T., Suhl, H. eds.), Vol. 1, p. 85, New York – London, Academic Press 1963
15. Kahn, O. in: Magneto-Structural Correlations in Exchange Coupled Systems (Willett, R. D., Gatteschi, D., Kahn, O. eds.), p. 37, Dordrecht, D. Reidel 1985
16. Anderson, P. W. in: Magnetism (Rado, G. T., Suhl, H. eds.), Vol. 1, p. 25, New York – London, Academic Press 1963
17. Hay, P. J., Thibeault, J. C., Hoffmann, R.: J. Am. Chem. Soc. *97,* 4884 (1975)
18. Kahn, O., Briat, B.: J. Chem. Soc. Faraday II, *72,* 268 (1976)
19. Kahn, O., Charlot, M. F.: Nouv. J. Chim. *4,* 567 (1980)
20. Girerd, J. J., Journaux, Y., Kahn, O.: Chem. Phys. Letters *82,* 534 (1981)
21. Van Kalkeren, G., Schmidt, W. W., Block, R.: Physica *97B,* 315 (1979)
22. Girerd, J. J., Charlot, M. F., Kahn, O.: Mol. Phys. *34,* 1063 (1977)
23. Girerd, J. J.: Thèse d'Etat, Université de Paris Sud 1982
24. Tola, P., Kahn, O., Chauvel, C., Coudanne, H.: Nouv. J. Chim. *1,* 467 (1977)
25. Moriya, T.: Phys. Rev. B *120,* 91 (1960)

26. Kanamori, J. in: Magnetism (Rado, G. T., Suhl, H. eds.), Vol. 1, p. 127, New York – London, Academic Press 1963
27. Gatteschi, D., Bencini, A. in: Magneto-Structural Correlations in Exchange Coupled Systems (Willett, R. D., Gatteschi, D., Kahn, O. eds.), p. 241, Dordrecht, D. Reidel 1985
28. Bencini, A., Gatteschi, D.: Mol. Phys. *47,* 161 (1982)
29. Banci, L., Bencini, A., Gatteschi, D.: J. Am. Chem. Soc. *105,* 761 (1983)
30. Charlot, M. F., Journaux, Y., Kahn, O., Bencini, A., Gatteschi, D., Zanchini, C.: Inorg. Chem. *25,* 1060 (1986)
31. Chao, C. C.: J. Magn. Reson. *10,* 1 (1973)
32. Buluggiu, E., Vera, A.: Z. Naturforsh. *31a,* 911 (1976)
33. Buluggiu, E.: J. Phys. Chem. Solids *41,* 1175 (1980)
34. Scaringe, R. P., Hodgson, D., Hatfield, W. E.: Mol. Phys. *35,* 701 (1978)
35. Griffith, J. S.: Struct. Bonding *10,* 87 (1972)
36. Kahn, O., Galy, J., Journaux, Y., Jaud, J., Morgenstern-Badarau, I.: J. Am. Chem. Soc. *104,* 2165 (1982)
37. Galy, J., Jaud, J., Kahn, O., Tola, P.: Inorg. Chim. Acta *36,* 229 (1979)
38. Julve, M., Verdaguer, M., Charlot, M. F., Kahn, O., Claude, R.: ibid. *82,* 5 (1984)
39. Kahn, O.: Comments Inorg. Chem. *3,* 105 (1984)
40. Crawford, W. H., Richardson, H. W., Wasson, J. R., Hodgson, D. J., Hatfield, W. E.: Inorg. Chem. *15,* 2107 (1976)
41. Landee, C. P., Willett, R. D.: ibid. *20,* 2521 (1981)
42. Ginsberg, A. P., Martin, R. L., Brookes, R. W., Sherwood, R. C.: ibid. *11,* 2884 (1972)
43. Duggan, D. M., Hendrickson, D. N.: ibid. *13,* 2929 (1974)
44. Laskowski, E. J., Felthouse, T. R., Hendrickson, D. N., Long, G.: ibid. *15,* 2908 (1976)
45. Young, K. O., O'Connor, C. J., Sinn, E., Carlin, R. L.: ibid. *18,* 804 (1979)
46. Journaux, Y., Kahn, O.: J. Chem. Soc. Dalton 1575 (1979)
47. Willett, R. D., Gaura, R. M., Landee, C. P. in: Extended Linear Chain Compounds (Miller, J. S. ed.), Vol. 3, p. 143, New York – London, Plenum Press 1983
48. Hatfield, W. E., Estes, W. E., Marsch, W. E., Pickens, M. W., Ten Haar, L. W., Weller,iR. R. ic: Extended Linear Chain Compounds (Miller, J. S. ed.), Vol. 3, p. 43, New York – London, Plenum Press 1983
49. Journaux, Y., Kahn, O., Zarembowitch, J., Galy, J., Jaud, J.: J. Am. Chem. Soc. *105,* 7585 (1983)
50. Carlin, R. L., Van Duyneveldt, A. J.: Magnetic Properties of Transition Metal Compounds, New York, Springer 1977
51. Lehninger, A. L.: The Mitrochondrion, New York, Benjamin 1973
52. Malström, B. G. in: Metal-Ion Activation of Dioxygen (Spiro, T. G. ed.), p. 181, New York, Wiley 1980
53. Cytochrome oxidase (King, T. E., Orii, Y., Chance, B., Okunuki, K. eds.), New York, Elsevier 1979
54. Palmer, G., Babcok, G. T., Vickery, L. E.: Proc. Natl. Acad. Sci. USA *73,* 2206 (1976)
55. Babcock, G. T., Vickery, L. E., Palmer, G.: J. Biol. Chem. *253,* 2400 (1978)
56. Tweedle, M. F., Wilson, L. J., Garcia-Iniguez, L., Babcock, G. T., Palmer, G.: ibid. *252,* 8065 (1978)
57. Moss, T. H., Shapiro, E., King, T. E., Beinert, H., Hartzell, C. R.: ibid. *253,* 8072 (1978)
58. Falk, K. E., Vänngard, T., Angström, J.: FEBS Lett. *75,* 23 (1977)
59. Chance, B., Powers, L.: Biophys. J. *33,* 95 (1981)
60. Powers, L., Chance, B., Ching, Y., Angiolillo, P.: ibid. *34,* 465 (1981)
61. Powers, L., Chance, B.: J. Inorg. Biochem. *23,* 207 (1985)
62. Scott, R. A., Schwartz, J. R., Cramer, S. P. in: Biological and Inorganic Copper Chemistry (Karlin, K. D., Zubieta, J. eds.), Vol. 1, p. 41, Guilderland, Adenine 1985
63. Reed, C. A., Landrum, J. T.: FEBS *106,* 265 (1979)
64. Petty, R. H., Welch, B. R., Wilson, L. J., Bottomley, L. A., Kadish, K. M.: J. Am. Chem. Soc. *102,* 611 (1980)
65. Berry, K. J., Clark, P. E., Gunter, M. J., Murray, K. S.: Nouv. J. Chim. *4,* 581 (1980)
66. Gunter, M. J., Mander, L. N., McLaughlin, G. M., Murray, K. S., Berry, K. J., Clark, P. E., Bucklingham, D. A.: J. Am. Chem. Soc. *102,* 1470 (1980)
67. Gunter, M. J., Mander, L. N., Murray, K. S., Clark, P. E.: ibid. *103,* 6784 (1981)

68. Lukas, B., Miller, J. R., Silver, J., Wilson, M. T., Morrison, I. E. G.: J. Chem. Soc. Dalton 1035 (1982)
69. Elliott, C. M., Akabori, K.: J. Am. Chem. Soc. *104,* 2671 (1982)
70. Dessens, S. E., Merrill, C. L., Saxton, R. J., Ilaria, Jr. R. L., Lindsey, J. W., Wilson, L. J.: ibid. *104,* 4357 (1982)
71. Saxton, R. J., Wilson, L. J.: J. Chem. Soc. Chem. Commun. 359 (1984)
72. Petty, R. H., Wilson, L. J.: ibid. 483 (1978)
73. Okawa, H., Kanda, W., Kida, S.: Chem. Letters 1281 (1980)
74. Brewer, G. A., Sinn, E.: Inorg. Chem. *23,* 2532 (1984)
75. Thomson, A. J., Johnson, M. K., Greenwood, C., Gooding, P. E.: Biochem. J. *193,* 687 (1981)
76. Thomson, A. J., Greenwood, C., Gadsby, P. M. A., Peterson, J., Eglinton, D. G., Hill, B. C., Nicholls, P.: J. Inorg. Biochem. *23,* 187 (1985)
77. Kent, T. A., Münck, E., Dunham, W. R., Filter, W. F., Findling, K. L., Yoshida, T., Fee, J. A.: J. Biol. Chem. *257,* 2489 (1982)
78. Gunter, M. J., Berry, K. J., Murray, K. S.: J. Am. Chem. Soc. *106,* 4227 (1984)
79. Okawa, H., Nishida, Y., Tanaka, M., Kida, S.: Bull. Chem. Soc. Jpn. *50,* 127 (1977)
80. Torihara, N., Okawa, H., Kida, S.: Inorg. Chim. Acta *26,* 97 (1976)
81. Morgenstern-Badarau, I., Rerat, M., Kahn, O., Jaud, J., Galy, J.: Inorg. Chem. *21,* 3050 (1982)
82. Desjardins, S., Morgenstern-Badarau, I., Kahn, O.: ibid. *23,* 3833 (1984)
83. Beale, J. P., Cunningham, J. A., Phillips, D. J.: Inorg. Chim. Acta *33,* 113 (1979)
84. Tanaka, M., Kitaoka, M., Okawa, H., Kida, S.: Bull. Chem. Soc. Jpn. *49,* 2469 (1976)
85. Mikuriya, M., Okawa, H., Kida, S., Ueda, I.: ibid. *51,* 2920 (1978)
86. Lambert, S. L., Spiro, C. L., Gagné, R. R., Hendrickson, D. N.: Inorg. Chem. *21,* 68 (1982)
87. Pilkington, N. H., Robson, R.: Aus. J. Chem. *23,* 2225 (1970)
88. O'Connor, C. J., Freyberg, D. P., Sinn, E.: Inorg. Chem. *18,* 1077 (1979)
89. Journaux, Y., Kahn, O., Morgenstern-Badarau, I., Galy, J., Jaud, J., Bencini, A., Gatteschi, D.: J. Am. Chem. Soc. *107,* 6305 (1985)
90. Bencini, A., Gatteschi, D.: Mol. Phys. *54,* 969 (1985)
91. Bencini, A., Caneschi, A., Dei, A., Gatteschi, D., Zanchini, C., Kahn, O.: Inorg. Chem. *25,* 1374 (1986)
92. Kokoszka, G. F., Allen, Jr. H. C., Gordon, G.: J. Chem. Phys.*46,* 3020 (1967)
93. Kokoszka, G. F., Duerst, R. W.: Coordin. Chem. Rev. *5,* 209 (1970)
94. Banci, L., Bencini, A., Gatteschi, D., Dei, A.: Inorg. Chim. Acta *36,* L419 (1979)
95. Banci, L., Bencini, A., Dei, A., Gatteschi, D.: Inorg. Chem. *20,* 393 (1981)
96. Banci, L., Bencini, A., Benelli, C., Dei, A., Gatteschi, D.: ibid. *20,* 1399 (1981)
97. Banci, L., Bencini, A., Benelli, C., Gatteschi, D.: ibid. *21,* 3868 (1982)
98. Bencini, A., Benelli, C., Gatteschi, D., Zanchini, C.: J. Am. Chem. Soc. *102,* 5820 (1980)
99. Krost, D. A., McPherson, G. L.: ibid. *100,* 987 (1978)
100. Paulson, J. A., Krost, D. A., McPherson, G. L., Rogers, R. D., Atwood, J. L.: Inorg. Chem. *19,* 2519 (1980)
101. McGarvey, B. R.: J. Phys. Chem. *71,* 51 (1966)
102. Abragam, A., Bleaney, B.: Electron Paramagnetic Resonance of Transition Ions, New York, Clarendon Press 1970
103. Gagné, R. R., Spiro, C. L., Smith, T. J., Hamann, C. A., Thies, W. R., Shiemke, A. K.: J. Am. Chem. Soc. *103,* 4073 (1981)
104. Eremin, M. V., Rakitin, Yu V.: Phys. Stat. Sol. b *80,* 579 (1977); *82,* 221 (1977); *85,* 783 (1978)
105. Eremin, M. V., Kalinenkov, V. N., Rakitin, Yu V.: ibid. *89,* 503 (1978); *90,* 123 (1978)
106. Morgenstern-Badarau, I., Cocco, D., Desideri, A., Rotilio, G., Jordanov, J., Dupré, N.: J. Am. Chem. Soc. *108,* 300 (1986)
107. Tainer, J. A., Getzoff, E. D., Beem, K. M., Richardson, J. S., Richardson, D. C.: J. Mol. Biol. *160,* 181 (1982)
108. Valentine, J. S., Pantoliano, M. W. in: Copper Proteins (Spiro, T. G. ed.), p. 292, New York, Wiley Interscience 1982
109. Hartmann-Boutron, F.: J. Phys. Paris *29,* 212 (1968)
110. Elliott, R. J., Thorpe, M. F.: J. Appl. Phys. *39,* 802 (1968)
111. Levy, P. M.: Phys. Rev. *117,* 509 (1969)

112. Lines, M. E.: J. Chem. Phys. *55,* 2977 (1971)
113. Lines, M. E., Ginsberg, A. P., Martin, R. L., Sherwood, R. C.: J. Chem. Phys. *57,* 1 (1972)
114. Barraclough, C. G., Gregson, A. K.: J. Chem. Soc. Faraday Trans. II 177 (1972)
115. Kahn, O.: Mol. Phys. *29,* 1039 (1975)
116. Stevens, K. W. H.: Phys. Rep. *24,* 1 (1976)
117. Drillon, M., Georges, R.: Phys. Rev. B *24,* 1278 (1981); *26,* 3882 (1982)
118. Briat, B., Kahn, O., Morgenstern-Badarau, I.: Inorg. Chem. *20,* 4193 (1981)
119. Fuchikami, N., Block, R.: Physica B + C *112,* 369 (1982)
120. Kahn, O., Tola, P., Coudanne, H.: Chem. Phys. *42,* 355 (1979)
121. Griffith, J. S.: The Theory of Transition Metal Ions, Cambridge, Cambridge University Press 1971
122. Bencini, A., Gatteschi, D., Zanchini, C.: Mol. Phys. *56,* 97 (1985)
123. Taylor, C. P. S.: Biochim. Biophys. Acta *491,* 137 (1977)
124. Palmer, G. in: The Porphyrins (Dolphin, D. ed.), Vol. 4, p. 313, New York, Academic Press 1979
125. Gruber, S. J., Harris, C. M., Sinn, E.: J. Chem. Phys. *49,* 2183 (1968)
126. Gruber, S. J., Harris, C. M., Sinn, E.: J. Inorg. Nucl. Chem. *30,* 1805 (1968)
127. Sinn, E. in: Biological and Inorganic Copper Chemistry (Karlin, K. D., Zubieta, J. eds.), Vol. 2, p. 195, Guilderland, Adenine 1985
128. Bencini, A., Benelli, C., Dei, A., Gatteschi, D.: Inorg. Chem. *24,* 695 (1985)
129. Morgenstern-Badarau, I., Wickman, H. H.: ibid. *24,* 1889 (1985)
130. Sekutowski, D., Jungst, R., Stucky, G. D.: ibid. *17,* 1848 (1978)
131. Journaux, Y., Sletten, J., Kahn, O.: ibid. *25,* 439 (1986)
132. Kahn, O.: Angew. Chem. Int. Ed. Engl. *24,* 834 (1985)
133. Verdaguer, M., Julve, M., Gleizes, A., Kahn, O.: Nouv. J. Chim. *9,* 325 (1985)
134. Bencini, A., Benelli, C., Caneschi, A., Carlin, R. L., Dei, A., Gatteschi, D.: J. Am. Chem. Soc. *107,* 8128 (1985)
135. Bencini, A., Benelli, C., Caneschi, A., Dei, A., Gatteschi, D.: Inorg. Chem. *25,* 572 (1986)
136. Ginsberg, A. P., Martin, R. L., Sherwood, R. C.: Inorg. Chem. *7,* 932 (1968)
137. Pei, Y., Journaux, Y., Kahn, O., Dei, A., Gatteschi, D.: J. Chem. Soc. Chem. Comm. 1300 (1986)
138. Gleizes, A., Verdaguer, M.: J. Am. Chem. Soc. *103,* 7373 (1981)
139. Verdaguer, M., Julve, M., Michalowicz, A., Kahn, O.: Inorg. Chem. *22,* 2624 (1983)
140. Orbach, R.: Phys. Rev. *115,* 1181 (1959)
141. Bonner, J. C., Fisher, M. E.: Phys. Rev. A *135,* 640 (1964)
142. Drillon, M., Gianduzzo, J. C., Georges, R.: Phys. Lett. *96A,* 413 (1983)
143. Coronado, E., Drillon, M., Beltran, D., Bernier, J. C.: Inorg. Chem. *23,* 4000 (1984)
144. Georges, R., Curely, J., Drillon, M.: J. Appl. Phys. *58,* 914 (1985)
145. Curely, J., Georges, R., Drillon, M.: Phys. Rev. B *33,* 6243 (1986)
146. Drillon, M., Coronado, E., Beltran, D., Georges, R.: J. Appl. Phys. *57,* 3353 (1984)
147. Drillon, M., Coronado, E., Beltran, D., Curely, J., Georges, R., Nugteren, P. R., De Jongh, L. J., Genicon, J. L.: J. Mag. Mag. Mater. *54–57,* 1507 (1986)
148. Drillon, M., Coronado, E., Beltran, D., Georges, R.: Chem. Phys. *79,* 449 (1983)
149. Fisher, M. E.: Am. J. Phys. *32,* 241 (1974)
150. Seiden, J.: J. Phys. (Paris) Lett. *44,* L947 (1983)
151. Verdaguer, M., Gleizes, A., Renard, J. P., Seiden, J.: Phys. Rev. B *29,* 5144 (1984)
152. Gleizes, A., Verdaguer, M.: J. Am. Chem. Soc. *106,* 3727 (1984)
153. Pei, Y., Sletten, J., Kahn, O.: ibid. *108,* 3143 (1986)
154. Pei. Y., Verdaguer, M., Kahn, O., Sletten, J., Renard, J. P.: Inorg. Chem. *26,* 138 (1987)
155. Pei, Y., Verdaguer, M., Kahn, O., Sletten, J., Renard, J. P.: J. Am. Chem. Soc. *108,* 7428 (1986)
156. Pozhidaev, A. I., Polynova, T. N., Porai-Koshits, M. A., Neronova, N. N.: Zh. Strukt. Khim. *14,* 570 (1973)
157. Coronado, E., Drillon, M., Fuertes, A., Beltran, D., Mosset, A., Galy, J.: J. Am. Chem. Soc. *108,* 900 (1986)
158. Beltran, D., Escriva, E., Drillon, M.: J. Chem. Soc., Faraday Trans. 2, *78,* 1773 (1982)
159. Seiden, J.: unpublished result
160. Peierls, R. E.: Quantum theory of Solids, London, Oxford Univ. Press 1955

161. Güdel, H. V. in: Magneto-Structural Correlations in Exchange Coupled Systems (Willett, R. D., Gatteschi, D., Kahn, O. eds.), p. 329, Dordrecht, D. Reidel 1985
162. Mathey, Y., Greig, D. R., Shriver, D. F.: Inorg. Chem. *21,* 3409 (1982)
163. Daudey, J. P., De Loth, P., Malrieu, J. P. in: Magneto-Structural Correlations in Exchange Coupled Systems (Willett, R. D., Gatteschi, D., Kahn, O. eds.), p. 87, Dordrecht, D. Reidel 1985
164. De Loth, P., Cassoux, P., Daudey, J. P., Malrieu, J. P.: J. Am. Chem. Soc. *103,* 4007 (1981)
165. Charlot, M. F., Verdaguer, M., Journaux, Y., De Loth, P., Daudey, J. P.: Inorg. Chem. *23,* 3802 (1984)
166. Iwamura, H., Sugawara, T., Itoh, K., Takui, T.: Mol. Cryst. Liquid Cryst. *125,* 251 (1985)
167. Sugawara, T., Bandow, S., Kimura, K., Iwamura, H., Itoh, K.: J. Am. Chem. Soc. *108,* 368 (1986)

Author Index Volumes 1–68

GPSR Compliance

The European Union's (EU) General Product Safety Regulation (GPSR) is a set of rules that requires consumer products to be safe and our obligations to ensure this.

If you have any concerns about our products, you can contact us on ProductSafety@springernature.com

In case Publisher is established outside the EU, the EU authorized representative is:

Springer Nature Customer Service Center GmbH
Europaplatz 3
69115 Heidelberg, Germany

Batch number: 10620881

Printed by Printforce, the Netherlands